# Management for Professionals

The Springer series *Management for Professionals* comprises high-level business and management books for executives. The authors are experienced business professionals and renowned professors who combine scientific background, best practice, and entrepreneurial vision to provide powerful insights into how to achieve business excellence.

More information about this series at http://www.springer.com/series/10101

Claude Diderich

# Design Thinking for Strategy

## Innovating Towards Competitive Advantage

 Springer

Claude Diderich
innovate.d llc
Richterswil, Switzerland

ISSN 2192-8096          ISSN 2192-810X   (electronic)
Management for Professionals
ISBN 978-3-030-25877-1          ISBN 978-3-030-25875-7   (eBook)
https://doi.org/10.1007/978-3-030-25875-7

This Springer imprint is published by the registered company Springer Nature Switzerland AG
The registered company address is: Gewerbestrasse 11, 6330 Cham, Switzerland

# Preface

The business environment is changing more rapidly over recent years than it has ever in the past. Traditional, deductive, data-driven strategy development approaches do not keep up with that pace of change. Too much time is spent on analyzing data, and too little time is used to understand customer needs and their jobs-to-be-done. Design thinking, initially used by architects and urban planners, has become a mainstream wicked problem-solving approach putting customers and their needs at the forefront.

This book describes how the design thinking approach can be used to develop and validate business strategies that are desirable (customers are interested in the offerings), feasible (firms can deliver upon the promises made with their offerings), viable (firms can generate a sustainable profit from its operations), and competitive (customers understand the differentiating value offered). It gives a hands-on approach to strategy that can be applied in both a start-up and a corporate environment.

First and foremost, I would like to thank everyone, especially many of the customers who have been working with me over the years and have helped me refine the design thinking-based strategy approach described in this book. I would also like to thank Anna Biamonte for her valuable input making this book more readable. A great thank goes to Christian Rauscher from Springer Verlag for accepting to take up this book project and all the staff for handling it. Another great thank goes to all my friends, including a special journalist from Geneva and her daughter, for having supported me while writing this book, and hopefully thereafter. Finally, I am very grateful to my parents for having taught me, by their example, to pursue one's dream however extreme it may look.

Richterswil, Switzerland
June 2019

Dr. Claude Diderich

# Contents

**Part I    The Concepts and Theories Behind Innovative Strategy
Design**

**1    Understanding the Need for a New Approach to Strategy
Development** ........................................................... 3
   1.1    Understanding the Concept of Strategy .................... 4
         1.1.1   Traditional Definitions of Strategy .............. 5
         1.1.2   Strategy from a Designer's Perspective ........... 6
         1.1.3   A Distinct Definition of Strategy................ 7
   1.2    Traditional Strategy Development Processes .............. 7
         1.2.1   Prescriptive School ......................... 7
         1.2.2   Descriptive School .......................... 8
   1.3    Challenges Faced by Traditional Approaches to Strategy
        Design............................................ 9
   1.4    Design Thinking as a Solution ...................... 10
         1.4.1   Design Thinking Approach..................... 11
         1.4.2   Delivering Value to Customers ................. 12
         1.4.3   A Common Language ....................... 12
         1.4.4   Integrating Stakeholders ...................... 13
         1.4.5   A Three Layers Process ...................... 13
   References ............................................. 13

**2    Recognizing Key Insights That Make Design Thinking Valuable
to Strategy** ........................................................ 15
   2.1    The Value of Design Thinking ...................... 15
         2.1.1   Customer-Centric Problem Solving .............. 17
         2.1.2   Iteratively Improving Through Prototyping
               and Validating ............................. 18
         2.1.3   Validating Ideas with Stakeholders .............. 19
         2.1.4   Combining Analytical Thinking and Intuition ....... 19
   2.2    A Look at the History of Design Thinking................ 20
         2.2.1   The 1970s ................................ 20
         2.2.2   The 1980s ................................ 21

|  | 2.2.3 | The 1990s | 22 |
|  | 2.2.4 | The New Millenial | 22 |
| 2.3 | Design Thinking for Strategy | | 25 |
| References | | | 27 |

**3   Revisiting the Business Model Canvas as a Common Language** . . .  29

| 3.1 | The Role of the Business Model in the Context of Strategy Design | | 30 |
| 3.2 | The Lightweight Business Model | | 31 |
|  | 3.2.1 | Rationale and Conceptual Details | 32 |
| 3.3 | The Detailed Business Model | | 35 |
|  | 3.3.1 | Rationale and Conceptual Details | 37 |
|  | 3.3.2 | Relations Between Elements of the Detailed Business Model | 42 |
| References | | | 44 |

**Part II   A Structured Approach to Strategy Development**

**4   Gaining a Collective Understanding of the Strategy Development Challenge** .  49

| 4.1 | Strategy Project Set-up | | 50 |
|  | 4.1.1 | Identifying Key Stakeholders and Their Roles | 50 |
|  | 4.1.2 | Fostering an Innovation Culture | 53 |
|  | 4.1.3 | Budget and Timeline | 54 |
|  | 4.1.4 | Assessment of the Change Capacity of and Underlying Risks for the Firm | 56 |
| 4.2 | Target Industry | | 59 |
|  | 4.2.1 | Incumbents | 59 |
|  | 4.2.2 | Mature Firms | 60 |
| 4.3 | Guiding Principles | | 61 |
| References | | | 62 |

**5   A Novel Strategy Development Process Based on Design Thinking** .  63

| 5.1 | Process Overview | | 64 |
| 5.2 | The Foundation Layer | | 64 |
|  | 5.2.1 | Strategy Brief | 66 |
|  | 5.2.2 | Understanding Today's Environment | 66 |
|  | 5.2.3 | Identifying Industry Trends | 67 |
|  | 5.2.4 | Choosing the Firm's Strategic Focus | 68 |
| 5.3 | The Business Model Layer | | 68 |
|  | 5.3.1 | Observing | 68 |
|  | 5.3.2 | Learning | 69 |
|  | 5.3.3 | Designing | 71 |
|  | 5.3.4 | Validating | 71 |

5.4 The Competition Layer . . . . . . . . . . . . . . . . . . . . . . . . . . . . . . 72
  5.4.1 Understanding the Competitive Landscape . . . . . . . . . . 73
  5.4.2 Communicating . . . . . . . . . . . . . . . . . . . . . . . . . . . . . . 74
References . . . . . . . . . . . . . . . . . . . . . . . . . . . . . . . . . . . . . . . . . . . . . . . 75

**Part III Laying the Foundation for a Successful Strategy**

**6 Understanding the Industry Environment and Its Implications
to Strategy** . . . . . . . . . . . . . . . . . . . . . . . . . . . . . . . . . . . . . . . . . . . . 79
 6.1 Current Environment Analysis . . . . . . . . . . . . . . . . . . . . . . . . . . 80
  6.1.1 Customers and Their Jobs-to-Be-Done . . . . . . . . . . . . 80
  6.1.2 Outsider Perspective on the Industry . . . . . . . . . . . . . . 83
  6.1.3 The Firm and Its Capabilities . . . . . . . . . . . . . . . . . . . 85
  6.1.4 Environmental Constraints . . . . . . . . . . . . . . . . . . . . . 87
 6.2 Industry Tends . . . . . . . . . . . . . . . . . . . . . . . . . . . . . . . . . . . . . . 90
  6.2.1 Customers . . . . . . . . . . . . . . . . . . . . . . . . . . . . . . . . . . 91
  6.2.2 Industry Structure . . . . . . . . . . . . . . . . . . . . . . . . . . . . 91
  6.2.3 Innovation and Technology . . . . . . . . . . . . . . . . . . . . 92
  6.2.4 Externalities . . . . . . . . . . . . . . . . . . . . . . . . . . . . . . . . 92
References . . . . . . . . . . . . . . . . . . . . . . . . . . . . . . . . . . . . . . . . . . . . . . . 92

**7 Choosing a Tangible Strategic Focus Rather Than Building
Upon an Abstract Vision** . . . . . . . . . . . . . . . . . . . . . . . . . . . . . . . . . 93
 7.1 Deriving the Strategic Focus Using Design Thinking . . . . . . . . 94
 7.2 Observing and Learning . . . . . . . . . . . . . . . . . . . . . . . . . . . . . . 96
 7.3 Designing Possible Strategic Focus Prototypes . . . . . . . . . . . . . 97
  7.3.1 Identifying Possible Strategic Focuses . . . . . . . . . . . . . 97
  7.3.2 Choosing How to Compete . . . . . . . . . . . . . . . . . . . . . 99
  7.3.3 Characteristics Supporting the Strategic Focus . . . . . . . 100
 7.4 Validating the Designed Strategic Focuses . . . . . . . . . . . . . . . . 103
  7.4.1 Checking for Consistency . . . . . . . . . . . . . . . . . . . . . . 103
  7.4.2 Formulating Strategy Hypothesis . . . . . . . . . . . . . . . . . 103
  7.4.3 Designing Strategy Experiments . . . . . . . . . . . . . . . . . 105
 7.5 Selecting the Target Strategic Focus . . . . . . . . . . . . . . . . . . . . . 106
References . . . . . . . . . . . . . . . . . . . . . . . . . . . . . . . . . . . . . . . . . . . . . . . 107

**Part IV Iteratively Developing the Business Model Underlying
the Strategy**

**8 Gaining Insights by Observing Target Customers in Their
Natural Environment** . . . . . . . . . . . . . . . . . . . . . . . . . . . . . . . . . . . . 111
 8.1 Observing Objectives . . . . . . . . . . . . . . . . . . . . . . . . . . . . . . . . 111
  8.1.1 Observing Mature Firms . . . . . . . . . . . . . . . . . . . . . . . 112
  8.1.2 Observing Start-up Firms . . . . . . . . . . . . . . . . . . . . . . 112
  8.1.3 Observing Disruptors . . . . . . . . . . . . . . . . . . . . . . . . . 113

8.2     Deriving Perspectives Based on the Strategic Focus . . . . . . . . .     114
8.3     The Observing Process . . . . . . . . . . . . . . . . . . . . . . . . . . . . . .     115
8.4     Identifying Target Populations . . . . . . . . . . . . . . . . . . . . . . . .     115
        8.4.1     Customers Strategic Focus-Based Target
                  Populations . . . . . . . . . . . . . . . . . . . . . . . . . . . . .     117
        8.4.2     Offerings Strategic Focus-Based Target
                  Populations . . . . . . . . . . . . . . . . . . . . . . . . . . . . .     118
        8.4.3     Capabilities Strategic Focus-Based Target
                  Populations . . . . . . . . . . . . . . . . . . . . . . . . . . . . .     118
        8.4.4     Financials Strategic Focus-Based Target
                  Populations . . . . . . . . . . . . . . . . . . . . . . . . . . . . .     119
8.5     Passively Observing . . . . . . . . . . . . . . . . . . . . . . . . . . . . . .     119
        8.5.1     Types of Observations . . . . . . . . . . . . . . . . . . . . . .     120
        8.5.2     Passively Observing Process . . . . . . . . . . . . . . . . . .     121
        8.5.3     Passive Observation Tools . . . . . . . . . . . . . . . . . . . .     122
8.6     Conducting Ethnographic Interviews . . . . . . . . . . . . . . . . . . .     124
8.7     Running Focus Groups . . . . . . . . . . . . . . . . . . . . . . . . . . . .     126
8.8     Performing Secondary Research . . . . . . . . . . . . . . . . . . . . . .     127
8.9     Timeline and Required Skills . . . . . . . . . . . . . . . . . . . . . . . .     128
References . . . . . . . . . . . . . . . . . . . . . . . . . . . . . . . . . . . . . . . . . .     129

9   **Understanding Target Populations and Their Jobs-to-Be-Done
    Through Learning** . . . . . . . . . . . . . . . . . . . . . . . . . . . . . . . . .     131
9.1     Learning Objectives . . . . . . . . . . . . . . . . . . . . . . . . . . . . . .     131
9.2     The Learning Process . . . . . . . . . . . . . . . . . . . . . . . . . . . . .     132
9.3     Selecting a Framework . . . . . . . . . . . . . . . . . . . . . . . . . . . .     132
        9.3.1     Understanding Customers . . . . . . . . . . . . . . . . . . . .     133
        9.3.2     Identifying Capabilities and Resources . . . . . . . . . . . .     135
        9.3.3     Comprehending Financials . . . . . . . . . . . . . . . . . . . .     136
9.4     Mapping and Clustering Insights to Gain Knowledge . . . . . . . .     137
9.5     Formulating and Validating Assumptions . . . . . . . . . . . . . . . . .     141
9.6     Timeline and Required Skills . . . . . . . . . . . . . . . . . . . . . . . .     142
References . . . . . . . . . . . . . . . . . . . . . . . . . . . . . . . . . . . . . . . . . .     143

10   **Shaping the Strategy by Designing Business Model Prototypes** . . .     145
10.1     Designing Objectives . . . . . . . . . . . . . . . . . . . . . . . . . . . . .     146
10.2     The Designing Process . . . . . . . . . . . . . . . . . . . . . . . . . . . .     147
10.3     Documenting the Current Detailed Business Model . . . . . . . . .     148
10.4     Generating Innovative Ideas . . . . . . . . . . . . . . . . . . . . . . . .     149
         10.4.1     Selecting a Target Population . . . . . . . . . . . . . . . . . .     149
         10.4.2     Ideation . . . . . . . . . . . . . . . . . . . . . . . . . . . . . . . .     150
         10.4.3     Typical Examples of Ideas . . . . . . . . . . . . . . . . . . . .     151
         10.4.4     Ideation Tools . . . . . . . . . . . . . . . . . . . . . . . . . . . .     156

10.5 Transforming Ideas into Business Model Prototypes . . . . . . . . . 159
10.6 Aggregating Prototypes Stemming from Multiple Ideas . . . . . . . 162
References . . . . . . . . . . . . . . . . . . . . . . . . . . . . . . . . . . . . . . . . . 163

**11 Managing Uncertainty Through Experiment-Based Validation** . . . . 165
11.1 Validating Objectives . . . . . . . . . . . . . . . . . . . . . . . . . . . . . 166
11.2 The Validating Process . . . . . . . . . . . . . . . . . . . . . . . . . . . . . 166
11.3 Formulating Assumptions . . . . . . . . . . . . . . . . . . . . . . . . . . . 168
11.4 Classifying and Prioritizing Assumptions . . . . . . . . . . . . . . . . 170
11.5 Designing and Conducting Experiments . . . . . . . . . . . . . . . . . . 171
      11.5.1 Typical Experiments . . . . . . . . . . . . . . . . . . . . . . . . 173
11.6 Validating Desirability, Viability, and Feasibility . . . . . . . . . . . 175
      11.6.1 Validating Desirability . . . . . . . . . . . . . . . . . . . . . . . 176
      11.6.2 Validating Viability . . . . . . . . . . . . . . . . . . . . . . . . . 177
      11.6.3 Validating Feasibility . . . . . . . . . . . . . . . . . . . . . . . . 177
11.7 Risks to Avoid . . . . . . . . . . . . . . . . . . . . . . . . . . . . . . . . . . 178
References . . . . . . . . . . . . . . . . . . . . . . . . . . . . . . . . . . . . . . . . . 178

**Part V Exposing the Designed Strategy to the Competitive Environment**

**12 Exploiting Findings from Game Theory to Succeed in a Competitive Environment** . . . . . . . . . . . . . . . . . . . . . . . . . 181
12.1 What Competitive Advantage Means . . . . . . . . . . . . . . . . . . . 181
12.2 Understanding How to Compete . . . . . . . . . . . . . . . . . . . . . . 182
      12.2.1 Competing on Differentiation or Uniqueness . . . . . . . . 183
      12.2.2 Competing by Being Superior . . . . . . . . . . . . . . . . . . . 184
      12.2.3 Handling Indifference . . . . . . . . . . . . . . . . . . . . . . . . 184
12.3 The Competing Process . . . . . . . . . . . . . . . . . . . . . . . . . . . . 185
12.4 The Competitive Landscape . . . . . . . . . . . . . . . . . . . . . . . . . 186
      12.4.1 Identifying Key Players . . . . . . . . . . . . . . . . . . . . . . . 186
      12.4.2 Possible Strategies for Competing . . . . . . . . . . . . . . . . 190
12.5 The Business Model in the Competitive Environment . . . . . . . . 191
12.6 Designing the Firm's Competitive Advantage . . . . . . . . . . . . . . 192
      12.6.1 Customers Based Competitive Advantage . . . . . . . . . . 193
      12.6.2 Offerings Based Competitive Advantage . . . . . . . . . . . 195
      12.6.3 Capabilities Based Competitive Advantage . . . . . . . . . 195
      12.6.4 Financials Based Competitive Advantage . . . . . . . . . . 195
12.7 Winning the Competition Game by Sustaining a Competitive Advantage Using Game Theory . . . . . . . . . . . . . . . . . . . . . . . 196
      12.7.1 Competitive Equilibrium . . . . . . . . . . . . . . . . . . . . . . 197
      12.7.2 Modeling Competition Using Game Trees . . . . . . . . . . 199
References . . . . . . . . . . . . . . . . . . . . . . . . . . . . . . . . . . . . . . . . . 200

**13  Laying the Groundwork for Strategy Implementation Through
      Stakeholder Focused Communication** ...................... 201
    13.1  The Communicating Process .......................... 202
    13.2  Understanding the Ground Rules ...................... 202
    13.3  Identifying the Audience ............................ 204
        13.3.1  Internal Audience ........................... 204
        13.3.2  External Audience .......................... 205
        13.3.3  Looking at the Audience from a Different
              Perspective .............................. 205
    13.4  Selecting Communication Channels .................... 206
        13.4.1  Face-to-Face Communication .................. 207
        13.4.2  Electronic Communication .................... 207
        13.4.3  Print Communication ........................ 208
    13.5  Laying-Out the Timeline ............................ 208
    13.6  Preparing the Message ............................. 209
        13.6.1  The Traditional Strategy Message ............... 210
        13.6.2  Crafting the Strategy Message in a Design Thinking
              World ................................... 210
    13.7  Telling the Story ................................. 212
    13.8  Validating that the Strategy Message is Understood ........ 214
    References ........................................ 214

**Index** ................................................ 215

# Part I
# The Concepts and Theories Behind Innovative Strategy Design

# Understanding the Need for a New Approach to Strategy Development

**1**

*The important thing is to not stop questioning. Curiosity has its own reason for existing*—Albert Einstein

Who still remembers Blockbuster? In the 1990s, Blockbuster was the market leader in movie rentals in the United States of America. It had a good understanding on how customers were renting movies, namely based on impulse. In addition, it had sound capabilities of renting videos to consumers through a large network of stores and generating revenues by charging a renting fee. Its strategy was developed using a traditional backward-looking analytical approach, resulting in a sound business model viable over many years. Still, in 2010 Blockbuster had to file for bankruptcy protection. So, what went wrong? The quick answer to that question is Netflix. But that is too short-sighted. Blockbuster failed to realize the changing environment and adjust its strategy accordingly.

Netflix took a different approach to strategy development (Shih and Kaufman 2014). Rather than relying on an analytical, backward looking methodology to strategy, it started by observing how customers rented movies and which pain-points they were faced with. They identified that a key pain-point faced by many movie renters was the late-fee charged by Blockbuster. Late-fees made-up a significant portion of the revenues in Blockbuster's business model. Netflix then tried to identify the causes of that late-fee pain. Why were customers faced with late fees? What were the reasons behind their inability to return the rented movies on time? And more importantly, how could this pain-point be addressed? The answer to the question was "lack of time to return the rented movies on time", which Netflix solved by offering a mail-based solution rather than an in-person solution.

Shipping large VHS cassettes was tedious and expensive. So, Netflix searched for an alternative. Although promising, live streaming via internet technology was not yet mature at that time. They looked for an alternative movie delivery medium and singled out DVDs as an emerging technology early 2000. Having solved the mail order size problem by replacing VHS cassettes with DVDs, Netflix was faced with another challenge. Not every household had a DVD player yet. Again, they

© Springer Nature Switzerland AG 2020
C. Diderich, *Design Thinking for Strategy*, Management for Professionals,
https://doi.org/10.1007/978-3-030-25875-7_1

adjusted their strategy by focusing on those customers who had recently bought a DVD player.

Another challenge inherited from the generic movie-rental business model was that, although there are many movies available at any given point in time, only a small number of movies, the blockbusters, are actively sought out and rented. This often led to blockbusters being unavailable for rental and customers being unhappy. Rather than increasing the number of blockbuster movie copies available, which would have been very costly, Netflix prototyped a different idea, trying to match movie availability with customer preferences. If a requested blockbuster movie was not available for renting, Netflix suggested a second-best alternative, based on an in-depth understanding of the customer's preferences. To do so, they developed a movie rating database and used pattern matching algorithms, that is, artificial intelligence, to identify potential movie alternatives. Iterative learning allowed refining the algorithm over time and resulted in the ability to optimize the movies to be held in stock.

A further issue Netflix faced in their mail order business was the delay introduced by mail delivery. Rather than going to a rental shop and returning with the desired movie, customers had to wait for the postal service to deliver the ordered movie DVD. To address that drawback, Netflix introduced an optimized hub-based supply chain management approach that sped up rented movies delivery. By thinking out of the box, they came up with a subscription-based pricing model relying on their capabilities to forecast customer movie preferences. This means, they rented out movies based on identified customer preferences without the customers having to place any order, making the overall process much more efficient.

Further down the road, Netflix introduced video-on-demand, and most recently concluded that it needed to produce its own content, like House of Cards, Orange is the New Black, Narcos, or The Crown, to be able to differentiate from competitors. And one can be curious to see what will be the next strategic adjustment that Netflix will make to address changing customer needs and technological innovations.

Most of the strategic choices made by Netflix would not have been possible using traditional analytical strategy development frameworks. Successful strategy design methods need to be able to cope with a rapidly chaining environment. They have to be forward-looking rather than backward-looking. They also require a superior understanding of customer needs, their felt pains and sought-after gains. Research-based, inwards looking, analytical approaches fail to cope with the dynamics of both. A paradigm shift is needed. Before describing a solution to the faced strategy development challenges, let me start by characterizing what strategy is and is not.

## 1.1  Understanding the Concept of Strategy

Through time, three complementary types of approaches to strategy have emerged: the environmental approaches, the capabilities- or resources-based approaches, and the customer-centric approaches. Figure 1.1 illustrates the three approaches and how they complement each-other.

**Fig. 1.1** Approaches to
business strategy focusing on
three complementary
elements

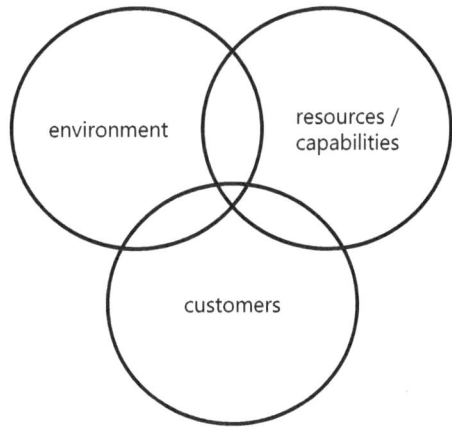

### 1.1.1  Traditional Definitions of Strategy

According to Porter (1985), strategy is about identifying and subsequently exploiting competitive advantages. Competitive advantage can either be achieved through cost leadership or through differentiation. More formally, developing a strategy means defining a particular configuration of the value chain, which is unique and sustainable over time, providing an offering that cannot easily be copied by competitors. Strategy is about choice, making trade-off decisions while competing (Porter 1996). Porter's definition of strategy is a combination of an environmental approach, such as his five forces model,[1] and a capability-based approach, for example, by focusing on the value chain concept. The process is analytical and focusing on convergent thinking.

Barney (1991, 2001a, 2001b) takes a different approach. He defines a strategy as a means of exploiting a firm's resources and related internal strengths to exploit environmental opportunities and neutralize external threats. The SWOT[2] analysis framework is at the core of developing such strategies. Success is based on effectively mapping resources to opportunities. Such strategies are called resource-based.

Mintzberg (1978, 1994, Mintzberg et al., 1988), another key strategy scholar, defines strategy as a stream of managerial decisions and actions, which are sometimes deliberate and at other times emergent. Strategic decisions are mostly based on managerial intuition and creativity, rather than analytical thinking. Mintzberg proposes a process-based approach, focusing on creativity and resulting in an integrated perspective of the firm.

---

[1]These five forces are (1) industry competition, (2) potential new entrants, (3) power of suppliers, (4) power of buyers, and (5) threat from substitute products and services.
[2]SWOT—Strength Weakness Opportunity Threat.

A commonality of these definitions of strategy is that they fail to include customers and their needs as a central element. Satisfying customer needs is seen as a consequence of strategic decisions rather than their driver.

## 1.1.2  Strategy from a Designer's Perspective

Traditional strategy development processes are analytical, linear, problem-focused, and backward-looking. They aim at exploiting the known by applying analytical and quantitative approaches. The analysis is often outsourced to consultants. By contrast, designers foster creativity, iterate, focus on solutions, and are forward looking. They aim at transforming existing conditions into new ones, to achieve future improvements. They approach problem solving from the point of view of the end-user and call for creative solutions by developing a deep understanding of unmet needs. Designers help structure team interactions by cultivating greater inclusiveness, empathy, and align individual goals around shared results (Mootee 2013). They put real people, not statistics, at the forefront. They emphasize the importance of exploration into the unknown, by focusing on qualitative and empathetic approaches. They engage stakeholders in co-creation. This makes their approach a sound alternative for strategy development.

The design thinking framework formalizes strategy development by offering a *strategy design process* supported by a common language addressing four key questions:

(1) What customer needs, pain-points, and sought-after gains are currently addressed or nor addressed, and what customers are not served?
(2) How can the identified needs and pain-points be addressed in a way that customers are willing to pay for?
(3) What are the distinct capabilities and resources required to achieve a sustainable competitive advantage in delivering upon the promises made, that is, addressing the identified needs?
(4) How is the strategy ensuring that sustainable profits can be generated?

In contrast to other approaches to strategy, design thinking focuses on *generating value for the customers in a differentiated and sustainable way*. Strategy is about choice, what to do and what not to do, whom to serve and whom not to serve. It is about competing in a given environment by differentiating from competitors and delivering superior value to customers. Strategy is the destination a firm aims at reaching, rather than the path to that destination. Strategy design, the identification of the destination, is separate from strategy implementation, the path towards the identified destination, as illustrated in Fig. 1.2. Strategy design is not about planning. It is even much less about developing a business case. Some of the largest companies have turned to design thinking as a way to deal with disruption and sustained competitiveness (Mootee 2013).

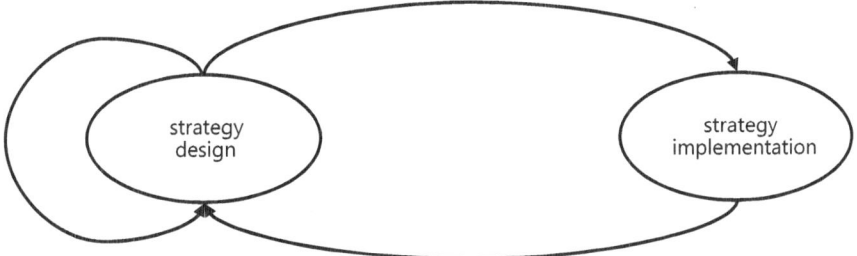

**Fig. 1.2** Iterating between strategy design and strategy implementation

### 1.1.3   A Distinct Definition of Strategy

Strategy in this book is defined as the combination of a *strategic focus*, that is, a differentiating value creator, a *business model* describing how the firms aims at delivering value to customers and other stakeholders, and an approach to differentiate, focusing on the *competitive positioning* of the firm in the business environment.

$$strategy = strategic\,focus + business\,model + competitive\,positioning$$

The strategic focus defines the big picture or the foundation. The business model considers how the firms creates and delivers value by addressing customer jobs-to-be-done relying on capabilities and resources and collaborating with partners and suppliers. Competitive positioning addresses the competitive environment and defines how the firm intends to use its competitive advantages to succeed. But how can we design such a strategy?

## 1.2   Traditional Strategy Development Processes

To better understand the challenges faced by applying traditional processes to design successful strategies, let me review the most prominent strategy development approaches and identify their strengths and weaknesses in a rapidly changing business environment. The academic literature on strategy broadly distinguishes between two types of strategy schools, the *prescriptive school* and the *descriptive school*.

### 1.2.1   Prescriptive School

The oldest prescriptive school is the *design school*, advocated by Chandler (1962), Ansoff (1965), and Andrews (1971). Note that the prescriptive strategy design school is unrelated to design thinking and must not be mixed up. The prescriptive design school focuses on matching internal capabilities to external opportunities. The core framework used is the SWOT analysis. Strategy development is the role of the firm's leader, the CEO or the chairman. Strategy design is separated from strategy implementation and kept simple and informal.

The second prescriptive school, the *planning school*, advocated by Steiner (1979), sees strategy development as an analytical and linearly convergent process. It mainly relies on strategic planning, which is how a firm's value chain is configured and resources are allocated, based on a set of strategic directions. In contrast with the design school, the planning school sees strategy development as a bottom-up approach involving line managers.

The third prescriptive school, mainly shaped by the work of Porter in his two landmarked books *Competitive Strategy* (Porter 1980) and *Competitive Advantage* (Porter 1985), is called the *positioning school*. It focuses on context, using frameworks such as the five forces model, rather than on process or on planning. It defines strategy as selecting from a constrained set of competitive positions and implementing the business logic behind them.

## 1.2.2   Descriptive School

The *descriptive school* towards strategy development places a higher value on the content rather than the process. It focuses on what the strategy represents rather than how it is derived. At least seven distinct descriptive schools can be identified.

The *cognitive school* defines strategy by looking at how people perceive patterns of data and process information. It focuses on what is happening in the mind of the strategy developer and how information is processed into insights.

The *entrepreneurial school* defines the strategy process as a visionary process that takes place in the head of a charismatic entrepreneur. The school stresses the innate nature of the key strategy development building blocks, that is, intuition, judgment, wisdom, experience, and insights. There exist three major sub-schools (Ott et al. 2017), those who strategize by doing—learning from experience, those who strategize by thinking—creating a holistic understanding, and those who strategize by iteratively doing and thinking.

Proponents of the *learning school* define strategy through what does and what does not work over time. They incorporate lessons learned into the overall strategy. The underlying principle of the learning school is that the world is too complex to allow a strategy to be developed all at once. Hence, the strategy of a firm emerges in small steps, as the firm's strategists learn.

Scholars stemming from the *political school* see strategy as the outcome of a negotiation process between powerhouses within the firm and with external stakeholders.

Strategy formulation in the *cultural school* is viewed as a fundamentally collective and cooperative process, involving various groups and departments within the firm. Strategy is seen as the outcome of a reflection on the corporate culture of the organization.

The *environmental school* defines strategy as the response to the challenges imposed by the external environment. The environment plays an active role in the strategy itself. It drives any strategic decision.

Last but not least, the *configurational school* defines strategy as the process of transforming an organization from one type of decision-making structure to another.

## 1.3 Challenges Faced by Traditional Approaches to Strategy Design

Four key challenges can be identified when trying to apply strategy development processes based on traditional strategy school thinking, whether prescriptive or descriptive, to the current fast-paced and ever-changing business environment:

(1) *Speed*—They are slow to execute.
   Traditional strategy development schools define sound approaches to the strategy development process. But they fail to cope with the fast-changing world, mainly due to their analytical foundations. They are slow, rigid, and often very ineffective.
(2) *Customer focus*—They tend not to focus on customers, their needs, their felt pains, and sought-after gains.
   Traditional strategy development approaches primarily focus on capabilities, those of the firm, those of competitors, and those defining the environment (suppliers, substitutes, etc.). They take an internal approach. They put the firm at the center of the strategy. But they fail to focus on customers and their jobs-to-be-done.
(3) *Complexity*—They are complex and hard to understand by the non-strategy trained manager or executive.
   Managers have a hard time navigating complex strategy frameworks, like Porter's five forces (Porter 1979), by themselves. It is an incorrect assumption to believe that successful managers are necessarily trained strategists.
(4) *Outsourcing*—More often than not, are large parts of the strategy development process outsourced to industry experts and strategy consultants.
   Consequently, the buy-in into the developed strategy is only half heated, resulting in a lack of follow-through.

Any good strategy development process requires guidance and simplicity, both in language and methodology. Successful strategy development requires decision makers to board a journey of discovery, exposing them to experiences that will align their beliefs with the outside world. The guidance and methodology may, and probably should, be facilitated by an independent method and moderation expert.

Yet the actual strategy design and validation work must be performed, or at least closely supervised, by those executives and directors who are ultimately responsible for its fate.

## 1.4  Design Thinking as a Solution

Any successful *strategy design process* addressing the identified challenges, should exhibit six key characteristics:

(1)  Consistent with the strategy design school, the strategy design process should be *top-down*, starting with designing and validation a sound foundation.
(2)  The strategy design process should follow an *agile*, just in time, sometimes also called lazy, approach, allowing for refinements and pivoting along the way.
(3)  The focus should be put on *designing the future* rather than analyzing the past, notwithstanding learning from historical successes and failures.
(4)  To ensure buy-in and subsequent successful implementation, the strategy design process should *integrate stakeholders* early in the design of the strategy, especially at the validation step.
(5)  There does not exist not a one size fits it all approach to strategy design. Any successful strategy design process must allow for different types of strategies, that is, *customer centric* strategies, *innovation-oriented* strategies, *capabilities-based* strategies, or *cost-driven* strategies.
(6)  And finally, the strategy design process must put the *targeted customers at the center* of any strategy design activity.

Design thinking is a method for solving wicked problems[3] (Churchman 1967), that is, problems with no upfront clear solution. It is based on abductive reasoning.[4] It aims at iteratively designing and validating solutions using a forward-looking approach and putting the customer at the center stage.

Strategy design is a typical wicked problem. It exhibits the four traits of openness, complexity, dynamism, and networking, as defined by Dorst (2015). The strategy design problem is an open problem, as its borders are unclear and permeable. There does not exist a single best solution. It is complex in the sense that it consists of many interrelated elements—like customers, competitors, suppliers, and regulators. Strategy solutions need to be dynamic, allowing to adapt to an ever faster changing environment. And solving the strategy design challenge requires

---

[3]A wicked problem is a problem that does not have a definite solution and as such cannot be solved using linear problem-solving techniques. Solving wicked problems requires continuous reformulation of the problem.
[4]Abductive reasoning is a form of logical inference which starts by observing, followed by searching for the simplest and most likely explanation, refining it until the solution is considered sound.

considering a network of stakeholders constantly influencing each other, rather than focusing on a single individual or group. Design thinking, combined with the business model framework and game theory, is predestined as a solid approach to developing sound strategies. In addition, integrating stakeholders throughout the strategy design process is key to success. Strategy design must become a mindset, rather than a procedural exercise (Bradley et al. 2011).

### 1.4.1  Design Thinking Approach

Design thinking is an abductive approach to problem solving, combining the advantages of design and thinking. It finds its roots in architectural and industrial design. The underlying process can be characterized by a two-by-two matrix, as shown in Fig. 1.3. The first dimension looks at the thinking process, which can be divergent or convergent. The second dimension describes the time period considered, which either focuses on the past or on the future.

In contrast with other approaches, design thinking zeroes in as much on the problem specification as it aims at finding a solution. It also moves away from identifying the single best solution, targeting superiority rather than optimality. Figure 1.3 illustrates the four steps that define design thinking, as it most appropriately applies to strategy. It summarizes the tools to be used during each of the four steps, namely observing, learning, designing, and validating. Different design thinking approaches use different terminologies for the various steps or decompose the activities in distinct ways, but the underlying philosophy remains the same.

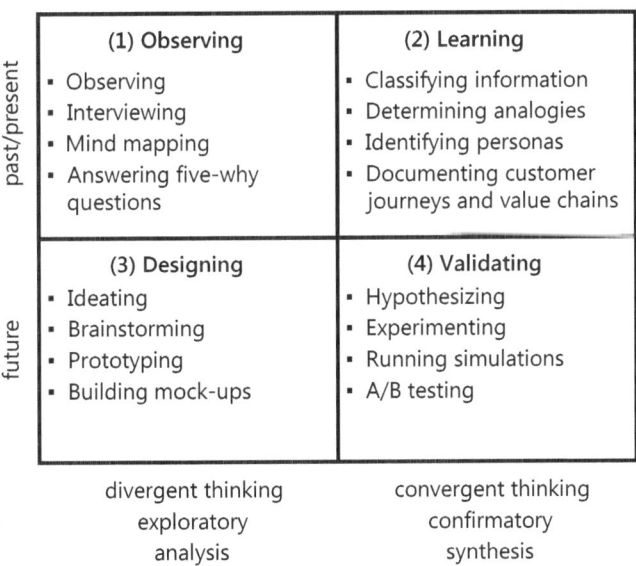

**Fig. 1.3** Four quadrants defining the design thinking approach, including possible tools to be used at each step

Chapter 2 explores design thinking in more detail and reviews different variations of design thinking processes from a historical perspective.

To avoid non-value-adding analysis, design thinking proceeds in an agile, just-in-time, way, moving to the next step as soon as enough insights have been gained. Whenever insights from a previous step turn out to be insufficient or incorrect, design thinking iterates back to the previous step and reconditions the missing or incorrect information. This allows proceeding in an agile way and avoids the use of unproductive labor whenever possible.

### 1.4.2  Delivering Value to Customers

Traditional strategy development processes primarily focus inwards on the firm and outwards on the competition, leaving customers as a residual. Design thinking supports building the strategy around the customers and their jobs-to-be-done. To be successful, strategy design must address four categories of questions related to customers (Brown 2009):

(1) *Desirable*—Are the offerings and associated value propositions underlying the strategy desired and sought-after by the targeted customers? Do they help satisfy a need, alleviate a pain, and/or provide additional gains to the targeted customers?
(2) *Feasible*—Can the firm deliver upon the promises made to the customers, both in terms of functionality and quality? Can the value proposition be upheld?
(3) *Viable*—Do customers consider the value of the offering worth paying for? Are customers willing to pay a price which will allow the firm to generate a profit?
(4) *Distinct*—Can customers distinguish the offering of the firm from that of its competitors? Do they value the uniqueness during their purchasing decision journey?

### 1.4.3  A Common Language

When individuals with diverse backgrounds, from marketing, product development, operations, legal and compliance, to finance, collaborate on the design of a new or the upgrade of an existing strategy, a common language is required. The business model canvas, introduced by Osterwalder and Pigneur (2010), provides such an easy to understand language, allowing for common fact finding, designing, and validating by all stakeholders involved in the strategy design process. Through its four major components, that is, customers, offerings, capabilities, and financials, it ensures a holistic approach to strategy design. Distinct levels of abstraction support the top-down approach.

### 1.4.4   Integrating Stakeholders

A strategy is only worth what senior management, executives, and members of the board of directors, believe it is. Having senior decision-makers on board is core to success. To achieve this needed buy-in, design thinking integrates all key stakeholders into the strategy design process from the beginning on. Senior managers are expected to participate, based on their experience, in the fact-finding steps (observing and learning steps). But more importantly, the designed strategy should be the outcome of a collaborative exercise between senior decision makers (designing step). Especially important is the active involvement of decision makers at the validation step. Participation in validating the assumptions ensures a higher degree of confidence and a commitment in the formulated strategy.

### 1.4.5   A Three Layers Process

The advocated strategy design process ensures success by decomposing strategy development into three layers, that is,

(1)   the *foundation layer*,
(2)   the *business model layer*, and
(3)   the *competition layer*.

Each layer focuses on a specific characteristic of a strategy, starting with an operationalized version of the vision concept—the foundation. Based on the foundation, the business model supporting the strategy is designed. It defines the key elements of a successful firm. The third layer focuses on competition and differentiation. It puts the business model into perspective and ensures a positioning that provides a lasting competitive advantage. Each layer is described and discussed in a separate part of this book, part III focusing on the foundation layer, part IV on the business model layer, and part V on the competition layer.

### References

Andrews, K. R. (1971). *The concepts of corporate strategy*. Homewood, IL: Irwin.
Ansoff, H. I. (1965). *Corporate strategy*. New York, NY: McGraw Hill.
Barney, J. B. (1991). Firm resources and sustained competitive advantage. *Journal of Management, 17*(1), 99–120.
Barney, J. B. (2001a). Resource-based theories of competitive advantage: A ten-year retrospective on the resource-based view. *Journal of Management, 27*(6), 643–650.
Barney, J. B. (2001b). Is the resource-based "view" a useful perspective for strategic management research? Yes. *Academy of Management Review, 26*(1), 41–56.

Bradley, C., Hirt, M., & Smit, S. (2011). Have you tested your strategy lately? *McKinsey Quarterly.*

Brown, T. (2009). *Change by design: How design thinking transforms organizations and inspires innovation.* New York, NY: HarperCollins Publishers.

Chandler, A. D. (1962). *Strategy and structure.* Cambridge, MA: MIT Press.

Churchman, C. W. (1967). Wicked problems. *Management Science, 4*(14), 141–142.

Dorst, K. (2015). *Frame innovation.* Cambridge, MA: MIT Press.

Mintzberg, H. (1978). Patterns in strategy formation. *Management Science, 24*(9), 934–948.

Mintzberg, H. (1994). *The rise and fall of strategic planning.* New York, NY: The Free Press.

Mintzberg, H., Lampel, J., Quinn, J. B., & Goshal, S. (1988). *The strategy process.* Upper Saddle River, NJ: Prentice Hall.

Mootee, I. (2013). *Design thinking for strategic innovation.* Hoboken, NJ: Wiley.

Osterwalder, A., & Pigneur, Y. (2010). *Business model generation.* Hoboken, NJ: Wiley.

Ott, T. E., Eisenhardt, K. M., & Bingham, C. B. (2017). Strategy formation in entrepreneurial settings: Past insights and future directions. *Strategic Entrepreneurship Journal, 11*(3), 306–325.

Porter, M. E. (1979). How competitive forces shape strategy. *Harvard Business Review, 57*(2), 137–145.

Porter, M. E. (1980). *Competitive strategy.* New York, NY: The Free Press.

Porter, M. E. (1985). *Competitive advantage.* New York, NY: The Free Press.

Porter, M. E. (1996). What is strategy? *Harvard Business Review, 74*(6), 61–78.

Shih, W., & Kaufman, S. (2014). *Netflix 2011. Case Study.* Boston, MA: Harvard Business School.

Steiner, G. (1979). *Strategic planning.* New York, NY: The Free Press.

# Recognizing Key Insights That Make Design Thinking Valuable to Strategy

**2**

*Never delegate understanding*—Charles Eames

In recent years, design thinking has become a buzzword for disruptive user-centered innovation. Its origins can be traced back to the early 1960s (Arnold 1959), namely to the participatory design movement that was characterized by software development. It was based on prototyping and incorporating customer feedback early in the development phase. Design thinking is a methodology, some call it a way of reasoning, some even an ecosystem (Diderich 2018), that combines logical thinking with creativity to understand the present and design the future. It starts by observing customers in their natural environment to learn their unmet needs, felt pains, sought-after gains, and jobs-to-be-done. Using ideation techniques combined with prototyping and experimentation, the gained insights are transformed into tested and viable solutions. Design thinking connects and integrates useful knowledge from arts and science alike, giving design a scientific basis (Buchanan 1992).

Design thinking relies on abductive reasoning as an effective way to alternate intuitive and deliberate actions. Abductive reasoning starts with a set of abstractions, that is, an incomplete set of observations, and seeks for the simplest and most likely solution. The initial solution is then improved upon through inference until it becomes a robust solution. Unlike deductive reasoning, abductive reasoning does not assume that the solution is contained in the premises of the problem. Quoting Einstein, "we cannot solve our problems with the same thinking we used when we created them".

## 2.1 The Value of Design Thinking

Design thinking addresses diverse shortcomings of analytical strategy development methods in a dynamic and fast-paced business environment. It aims at learning from methodologies used by designers, such as architects, artists, or creative directors, to

© Springer Nature Switzerland AG 2020
C. Diderich, *Design Thinking for Strategy*, Management for Professionals,
https://doi.org/10.1007/978-3-030-25875-7_2

solve problems which are incomplete by nature and cannot be solved by traditional linear problem-solving approaches.

Design thinking exhibits four key traits valuable to strategy design:

(1) Design thinking is *customer-centric*. Problem solving starts with observing and understanding customers and their needs, their suffered pains, their sought-after gains, and their jobs-to-be-done. Insights are acquired by focusing on observing and listening to customers in their natural environment, avoiding any interference that could distort the observed.

(2) Design thinking is *iterative* in nature. It incrementally addresses challenges, improving solutions step by step, considering what has previously been learned, and using resources (time and money) wisely. It allows avoiding unfocused data gathering and analysis.

(3) Design thinking is based on *prototyping and validating ideas*. It ensures that the designed solutions work. It does not assume that there exists a single best solution, but rather uses prototyping to identify trade-offs, validating them, and retaining those solutions that work.

(4) Design thinking combines the best of the two worlds of analytical and intuitive thinking, resulting in a so-called *abductive reasoning* approach.

Table 2.1 summarizes the four key traits of design thinking and explains their value to strategy development.

**Table 2.1** Key design thinking traits and their value

| | Design thinking trait | Value to strategy development |
|---|---|---|
| (1) | *Customer-centric approach*, putting customers at the forefront | – Ensuring customer needs are identified and met, their pains addressed, sought-after gains provided, and their jobs get done<br>– Creating unique and appreciated added value for customers<br>– Securing a willingness to pay |
| (2) | *Iterative process*, based on observing, learning, designing, and validating, supported by divergent and convergent thinking | – Well-defined systematic process leading to validated results<br>– Focused approach avoiding non-value-adding data gathering and analysis<br>– Agile, just-in-time, process due to its iterative nature |
| (3) | *Prototyped options*, designed and validated jointly with stakeholders | – Ensuring that the designed strategic options are aligned with stakeholder expectations<br>– Ascertaining that identified needs are met |
| (4) | Approach combining *analytical and intuitive thinking*, focusing on those insights that matter most | – Conscious use of resources (time and money)<br>– Constantly (re-)aligning efforts with set priorities<br>– Following "fail fast to succeed faster" philosophy by learning early from mistakes |

Design thinking is a systematic process for wicked problem solving as well as a visual language for communicating about ideas. Through its structure, design thinking ensures that resulting solutions generate value for the customers for whom they have been designed. By being iterative in nature, design thinking aims at solving 80% of the problem with 20% of the resources. This is achieved by reducing the complexity early on during the problem-solving process, by iteratively observing, learning, designing, and validating. Non-value-adding and time-consuming data gathering, and analysis steps are avoided whenever possible. Design thinking works best if the problem to be solved is poorly understood, there does not exist a single best solution, and it is impossible to layout a linear problem-solving process beforehand.

I illustrate the four traits of design thinking with examples, either from real life, or hypothetical, forward looking situations. Readers should keep in mind that these examples are not meant to be backwards looking case-studies. Their goal is to help understand the concepts, frameworks, and tools introduced. They should spur readers into thinking and coming-up with their own ideas. They offer a possible basis to formulate novel ideas or combine existing insights in a novel way. Since strategy is about being different and unique, following successful case studies does not help achieve that goal. I therefore avoid presenting exhaustive case studies.

## 2.1.1   Customer-Centric Problem Solving

Design thinking is based on the observation that solving typical business problems requires an in-depth understanding of the customers, their needs, their perceived pains, and their thought-after gains. Traditional analytical approaches rely on historical data, like surveys or past experiences, to understand customers and their needs. They put the focus on known facts from the past subsumed in data, answering the "what do customers need", rather than the "why do customers have specific needs" question. The rationale behind the data is often missed.

Rather than ask the customers what they want, as done by traditional customer and market research, design thinking investigates what customers do or do not do and why, what their jobs-to-be-done are. As Henry Ford is often quoted saying, if he had asked what customers want, they would have said, faster horses.[1] In contrast, observing customers and their behavior, the design thinking expert would have found out that the customer need or job-to-be-done is getting from point A to point B in a fast way without sacrificing flexibility and simplicity. By relying on intuition and experimenting jointly with customers in different environments, design thinking provides more relevant insights. It focuses on the unknown rather than the knowledgeable.

---

[1]According to Vlaskovits (2011) there is no evidence that Ford actually said that quote. However, even if he did not verbalize his thought on the apparent inability of customers to communicate their unmet needs, history indicates that Ford most certainly did think along those lines.

**Example** A relocation company was faced with shrinking margins due to online competition. Rather than accepting competition and focusing on cost cutting, the firm decided to identify customer segments that do not primarily buy on price and better understand their valued needs. To do so, it took a dual approach. On one hand, it conducted interviews with identified non-customers that chose competitors over the firm to find out what they valued and missed from the solution they retained. On the other hand, it conducted on-line social media research to identify what customers were praising and what they were complaining about.

Contrary to what the firm had initially thought, the identified customers, although being price-sensitive, where not solely buying on price, but also on quality of service. Also, customers showed more flexibility with respect to the relocation timeline than expected, within certain limits. More importantly, the focused analysis showed that expedite problem handling was highly valued, for instance when something broke or got lost during relocation, ideally through a dedicated contact, rather than an anonymous call center. Another key finding was the need for transparency along the whole customer journey, from searching a trustworthy relocation company, through understanding the services and options offered, to the final delivery. These insights allowed the firm to re-state its strategy, focusing on targeted customer segments rather than serving everyone, and considering the specific needs of those customers, in a way that allowed them to regain profitability.

## 2.1.2  Iteratively Improving Through Prototyping and Validating

Design thinking is based on the observation that it is not possible to get the solution of a wicked problem right the first time. Design thinking relies on iteratively trying out different options and improving solutions over time by considering what has been learned, what worked, and what did not work. In that sense, design thinking borrows ideas from agile, or just-in-time, methodologies and puts them into the context of ideation. In addition, stakeholders are actively involved in ideation, designing prototypes, and experimentation. Observations and insights are transformed into prototypes of ideas that can be validated with real customers. Each validation round leads to new observations and insights which allow improving upon previous prototypes. Successful design thinkers embrace the back and forth nature, making mistakes, learning from mistakes, and improving upon them, while knowing when good is good enough.

**Example** Having worked in a hospital, a team of students had identified an interesting challenge with pulse oximetry equipment: the wirings proved difficult to handle for the staff and hindered patient mobility. So, they came up with a wireless pulse oximetry prototype. They showed it to nurses to validate their idea, who immediately loved it. But when they talked to hospital administrators, who oversaw procurement, they were confronted with a "no interest in spending money on wireless pulse oximetry" answer, as administrators did not see the value of the solution. This lead the team of students to iterate and look for other applications of their idea. They identified the issue of infants dying from respiratory failure as a possible problem that could be solved with their wireless pulse oximetry system. Further iterations lead to an innovative solution, a sock solution that comfortably fits the equipment on infants and new-born babies. The OwletCare Baby Monitor was successfully launched in the U.S. market.

Coming back to Henry Ford's "faster horses challenge", a possible prototype could have been a carriage solution with multiple horses, but it would have missed the simplicity requirement. Alternatively, prototyping the idea of small individual trains could have come up. Again, validation would probably have failed on the need for tracks, rather than roads. Using the insights gained, the solution of a trackless train, is then not far away. Only the engine problem still needed addressing, getting from steam engines to combustion engines.

### 2.1.3   Validating Ideas with Stakeholders

Designed solutions are only good if deemed so by their actual stakeholders. Design thinking requires involving different stakeholders, especially those involved in decision making, into the validation of the designed prototypes. Depending on their skills, they are requested to perform validating experiments themselves. This allows them gaining first-hand experience and thus strengthens their confidence in the obtained results. Although decision makers are often reluctant to actively participate in assumption validation, they regularly value the insights gained ex-post. Unwillingness of decision makers to participate in assumption validation is often indicative of a reluctance to change. Being able to identify and address that reluctance at an early stage increases the probability of success.

> **Example** While developing a new business model for a multi-family office, the design team was confronted with the challenge of choosing the right pricing model, that is, relying on fixed prices, effort-based pricing, asset-based pricing, etc. As the team knew that this decision would be critical to success, not only with respect to customers embracing the offerings, but also to get buy-in from the executive team, they decided to involve key executives in finding out what pricing model is considered most appropriate by the targeted customers. To do so, they looked for executives willing to interview customers themselves (unfortunately not all found this a good idea) and coached them to do so. The outcomes from the interviews where not only that the executives identified the most appropriate pricing model to implement, it also strengthened their buy-in for the chosen model, as they had heard first-hand how customers think about price models and what they value, and thus no longer had to be convinced by a subordinated design team.

### 2.1.4   Combining Analytical Thinking and Intuition

Analytical thinking is based on using data combined with theoretical models and deriving insights to make sensible decisions. In today's world of big data, analytical thinking is often the preferred approach. It proceeds by understanding complex problems and decomposing them into simpler ones. To do so, analytical thinking starts with often unfocused data gathering and fact finding, followed by explicit search for matching patterns. Only at a later stage are the insights gained from the information combined to derive a solution, usually aiming directly for the optimal one (dashed line in Fig. 2.1).

Intuition, on the other hand, is based on the ability to acquire insights without significant amounts of data, evidence, or formal proofs (dotted line in Fig. 2.1). Intuition relies on unconscious pattern-recognition and instinct. Experience plays an important role in feeding the unconscious cognition, inner sensing process. Intuition often solves problems without being able to explain why, that is, validating the proposed results.

Design thinking aims a combining the advantages of the two extreme deductive and inductive problem-solving approaches into one method. The resulting abductive reasoning[2] framework underlying design thinking starts with observing to seek an initial simple and intuitive solution. Sometimes, the initial solution is only a partial solution or only partially addresses the problem at hand. Often, the initial solution results in a rephrased problem statement. By subsequently analyzing the intuitive solution and gathering data to validate or invalidate it, the initial solution is revised and improved upon. Design thinking uses abductive reasoning to infer ever improving solutions, up to the point where the outcome is considered good enough or can no longer be improved upon.

## 2.2   A Look at the History of Design Thinking

While ideas around the concept of industrial design can be traced back to the late 1940s and early 1950s, the concept of design thinking emerged for the first time in the 1960s in the context of participatory design (Arnold 1959). Participatory design was a movement characterized by quick software prototype development cycles, incorporating customer feedback into the prototyping process.

### 2.2.1   The 1970s

It is fair to say that the first milestone in the design thinking history was set by the publication of Herbert A. Simon's book *The Science of the Artificial* in 1968 (Simon 1968). He introduced a three-step process to solve complex decision problems:

---

[2]Abductive reasoning is a form of logical inference which starts with an observation then seeks to find the simplest and most likely explanation. It has been developed by the philosopher Charles Sander Pierce, who defended that no new idea could be developed by deduction or induction using past date (Martin 2009). One can understand abductive reasoning as inference to the best explanation.

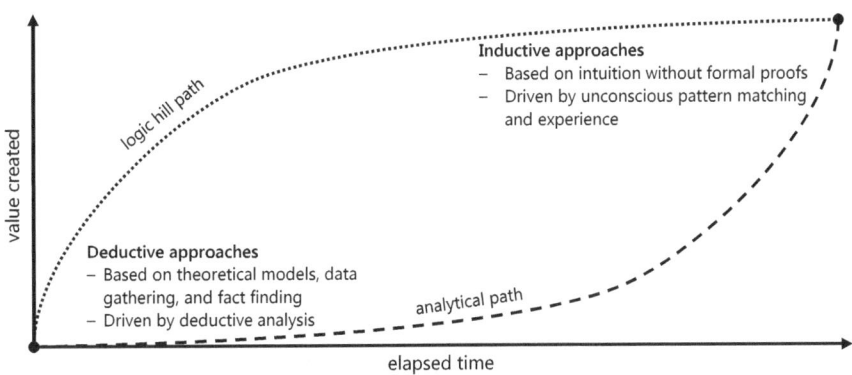

**Fig. 2.1** Deductive and inductive problem solving versus value created

(1)  intelligence gathering,
(2)  designing possible solutions, and
(3)  choosing a particular solution.

Compared with today's design thinking processes, Simon's approach was still linear in nature and did not put a strong focus on testing and validating designed solutions with customers. Non-linear problem solving was developed shortly thereafter, mostly by Koberg and Bagnall (1972).

End users were put at the center of software development design by Arnheim (1969) in a book called *Visual Thinking*. In 1973, McKim, professor in mechanical engineering at Stanford University and founder of the Stanford joint program on design, followed up on Arnheim's work publishing a book entitled *Experiences in Visual Thinking* (McKim 1973), elaborating how visual thinking can be used to successfully solve wicked problems.

## 2.2.2   The 1980s

The term Design Thinking, written in capital letters, describing a methodology of creative problem solving was introduced by Lawson (1980) in his seminal book *How Designers Think*. He described how the concept of design is used in architecture to solve problems. Architects, when compared to scientists, are more inclined to develop series of solutions until they find one that meets their criteria of being acceptable, rather than aim for the best possible solution right from the beginning, and therefore differs from the more linear process used by scientists and engineers.

In 1982, Cross published a paper titled *Designerly ways of knowing* (Cross 1982), that established some of the intrinsic characteristics underlying today's understanding of design thinking, namely a base for a coherent discipline of study and a focus on a broad audience. He noted that design thinking focuses on the future, creating new solutions, rather than on the past, elaborating on existing solutions. At its heart lies a visual language of modeling. Design thinking is viewed as one of three so-called cultures for representing and accessing human knowledge. These are:

(1) *Science culture*—Analytical, based on controlled experiments, relying on classification, and focusing on the physical world.
(2) *Humanities culture*—Analogy and metaphor based, focusing on evaluation and criticism, and driven by human experience.
(3) *Design culture*—Modeling driven, based on pattern formation and recognition, synthesis focused, and based on a man-made world.

A central feature of design thinking is generating satisfactory solutions fairly quickly rather than relying on prolonged problem analysis (Cross 1982, 2006, 2011). This characteristic is necessary to solve ill-defined and ill-structured wicked problems that do not have a single correct solution which can be found by exhaustive search. Solutions must be constructed, synthesized, rather than found, and recognized by the designer's own effort. Consequently, design thinking is largely based on tacit knowledge and difficult to externalize.

In 1987, Rowe published *Design Thinking* (Rowe 1987), describing methods and approaches used by architects and urban planners to solve wicked problems. University of Stanford's Faste, expanding on McKim's work, introduced design thinking as a method for teaching creative actions.

### 2.2.3  The 1990s

The 1990s were characterized by the adaption of design thinking to solving business problems. In 1991, Faste's colleagues Kelley, Moggridge, and Nuttall founded IDEO, a consulting company based on design thinking. IDEO was, and probably still is, the most prominent product and industrial design company embracing and advancing design thinking. Buchanan broadened the view on design thinking as a methodology for solving wicked problems in his paper called *Wicked Problems in Design Thinking* (Buchanan 1992).

### 2.2.4  The New Millenial

The new millennial was shaped by the development and introduction of formal processes to apply design thinking to problem solving. A large body of knowledge around design thinking, both from an academic and a practical perspective, has been developed and published over the years. The approaches described in this section cover the most relevant insights gained over time.

In 2001, the team led by Brown at IDEO, introduced its three-step process around inspiring, ideating, and implementing (Brown 2009).

In 2005, researchers at the newly founded d.School at Stanford University developed a five-step design thinking process that has been at the heart of many subsequent researches on design thinking processes. The five steps are:

(1) *Empathize*. This first step is about understanding the problem at hand. Observations, interviews, and measurements are some of the key tools used to gaining an objective, non-judgmental view of the challenge at hand. Key are empathy and customer-centricity.
(2) *Define*. During the second step of the design thinking process, the gained data is used to clearly define the problem at hand and describe the core challenge to solve in an objective way. The problem is defined in terms of customer and their needs, rather than the firm's internal goals. Sometimes the define step is compared to a root cause analysis taking a customer-centric perspective and using as input the data from the empathize step.
(3) *Ideate*. New possible solutions are created by starting with a large number of ideas and narrowing them down through eliminating those ideas that are unacceptable in terms of cost, value, time, resources, etc. More often than not does the ideation step include brainstorming or brain walking exercises.
(4) *Prototype*. Prototyping is about transforming ideas into actionable concepts that can be shared, reviewed, and validated. Prototypes need not be perfect and are iteratively refined and improved until they can demonstrate value from a customer perspective. At this step, several prototypes are usually defined.
(5) *Test*. Before selecting a prototype as the problem's solution, they are tested and validated. To do so, experiments are designed and performed. Based on the outcomes of the experiments, the prototypes are iteratively refined until a validated working solution is found.

The British design council introduced in 2005 its double diamond method. It is based on two iterations of divergent and convergent thinking steps. First, during the divergent discovery step, insights related to the problem at hand are collected. This step is similar to the empathize step of the d.School process. Next, applying convergent thinking, the problem to be solved is defined, as does the step with the same name in the d.School approach. During the third step, the develop step, divergent thinking is used to develop possible solution to the identified problems. It includes ideation and prototyping, as well as experimenting. Finally, convergent thinking is used to select the retained solution in the fourth step called deliver. The double diamond process can be considered a simplification of the Stanford d.School process. In addition, it can be perceived as adding a second dimension to the design thinking process, notably the amount of insights gained over time through divergent and convergent thinking.

The Hasso Plattner Institute of the University of Potsdam introduced in 2007 a design thinking process similar to the d.School one, based on six the six phases: (1) understand, (2) observe, (3) ideate, (4) prototype, (5) test, and (6) implement. The front-end of the process was slightly adjusted and an additional implementation step added. It remains an open question whether problem specification as well as implementation of the designed solution should be an integral part of design thinking or not.

Eppinger and Ulrich (1995) from the MIT, known for their research on product development, introduced a more analytical version of the design thinking process. It consists of the four steps (1) understand the problem, (2) develop possible solutions, (3) prototype, test, and refine the developed solutions, and (4) implement the retained solution. Although similar to other design thinking processes on paper, Eppinger's approach is rooted in an analytical engineering-based way of thinking. In addition, the iterative nature of design thinking is used in the third step, around prototyping, testing, and refining.

Schneider and Stickdorn (2011) introduced a design thinking process specifically tailored to service design. It encompasses four phases, that is, (1) explore, (2) create, (3) reflect, and (4) implement. Rather than rely on physical prototypes that can be tested, they suggest a mental approach replacing formal testing by a reflection phase.

In 2011, Liedtka and Ogilvie (2011) at the Darden School of the University of Virginia, introduced a variation of the design thinking process whereby they rephrased the different process steps as questions and combined them with activities and tools supporting answering the questions. The explicit goal of their approach focuses on solving wicked problems, rather than on generic design. The four questions to be answered are:

(1) *What is?* Answering the first question sets the scene for solving the considered wicked problem. It ensures that the real problem or opportunity to be tackled is correctly identified and well understood. Answers to the "what is" question, summarized in so-called design criteria, help avoid framing the problem to widely or too narrowly. Key tools supporting answering the "what is" questions, are personas, the customer journey mapping tool, the value chain analysis, as well as generic mind mapping frameworks.
(2) *What if?* Creativity starts with identifying possible opportunities which may solve the problem at hand. Answering the "what if" question directs the search for solutions to possible opportunities and avoids focusing on constraints. Key tools to generate ideas and options for solving the posed challenge are classical brainstorming as well as concept development.
(3) *What wows?* The third question focuses on evaluating possible solutions identified, focusing on retaining those that may be most relevant and valuable. As design thinking is essentially hypothesis driven, answering the "what wows" question requires formulating and validating assumptions behind the developed ideas. Rapid prototyping and assumption testing are key activities that help answer the "what wows" question.

(4) *What works?* Finally, the fourth question allows learning from the real world how the retained solution would be received and what value customers would see in it. Two key tools supporting finding out what works are customer co-creation as well as learning launch.

## 2.3 Design Thinking for Strategy

The simplest model of a creative process is a two-step process, an expanding stage of divergent thinking where many possibilities are generated, followed by convergent thinking, trending towards the best idea (Sawyer 2012).

Experience has shown that, out of the numerous design thinking processes developed so far, a variation of the double-diamond method is the one that works best to support strategy development. Indeed, it relates to the two phases of strategy development, understanding the past by looking backward, and designing the future by looking forward. In each phase, divergent thinking is followed by convergent thinking as illustrated in Fig. 2.2. There is no need to include a dedicated problem specification step in the process, as the strategy design challenge is well defined at the outset. In strategy work, the distinction between ideation and prototyping is also less relevant and often slippery, thus separating ideation from prototyping is not necessary. In addition, strategy implementation is best handled outside the strategy development process, as it requires a distinct skill set.

Each of the four steps of a design thinking-based strategy design process has a well-defined outcome: insights, knowledge, prototyped ideas, and validated strategy. This optimizes resource allocation and focuses on required skills. It is that focus, combined with categorized outcomes, that ensures process efficiency without losing creativity (Tschimmel 2012). Through proceeding iteratively, only adding

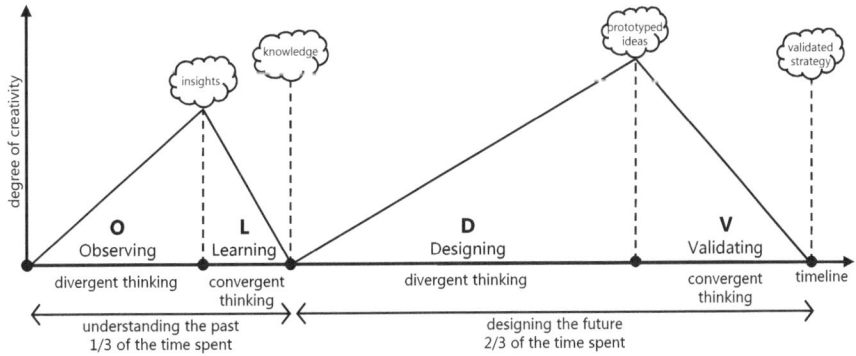

© Dr. Claude Diderich. Used with permission. Based on illustration from Diderich (2018)

**Fig. 2.2** Linearized version of the design thinking process used for strategy design

value analysis are performed, exploratory analysis during the divergent thinking steps (observing, designing), and confirmatory analysis during the convergent thinking steps (learning, validating).

I use the term *design thinking for strategy* ("DTS") to describe the design thinking approach illustrated in Fig. 2.2 to solving the strategy development challenge, whether for developing new strategies or improving upon existing ones. DTS extends beyond the traditional customer-centric way of thinking. DTS is strategy focus-centered design, whereby typical design thinking is customer-centered and can be seen as a special case of DTS, in which the strategy focus is set to be the customer. For example, a firm may want to become more effective by sharing production processes among multiple independent offerings. Such a strategy focus can be developed using DTS, without primarily being customer-centric. Although the customer still retains a key role in DTS, the focus is on exploiting a firm's invention capabilities, leveraging its core competencies, or generating value through financial engineering, just to name a few possibilities.

The four steps of the DTS process are:

(1) *Observing (divergent thinking, focusing on the past).* During the first step of the DTS process, the observing step, relevant insights are gathered. The goal is not to get an exact replication of the real world, but a first proxy that allows moving forward in the strategy design process. Observing aims at gaining insights without interfering, that is, changing the observed because of the way observations are conducted.

(2) *Learning (convergent thinking, focusing on the past).* During the learning step, the insights gained from observing are processed, clustered, synthesized, and transformed into knowledge. What is important is separated from what is not important. If, during the learning step, apparent information is considered missing, the process loops back to the observing step to gather the missing information.

(3) *Designing (divergent thinking, focusing on the future).* Next, during the designing step of the DTS process, ideas are generated, based on the learned knowledge, and transformed into prototypes. Prototypes may be either physical, like using the LEGO® SERIOUS PLAY® method, or mental models, like storyboards. They do not have to be complete. The goal is to build a representation of the proposed strategy that is realistic enough, so that it can be validated.

(4) *Validating (convergent thinking, focusing on the future).* During the fourth step of the DTS process, the validating step, the chosen strategy is tested by designing and performing experiments. This helps remove grid-locked discussions, often encountered in conference rooms (Liedtka et al. 2017). The goal of any experiment is attempting to identify potential weaknesses in the design made, rather than to prove its validity. If any of the experiments fails, the process reverts to the designing step where alternate ideas are prototyped and subsequently validated. This usually requires reviewing and potentially

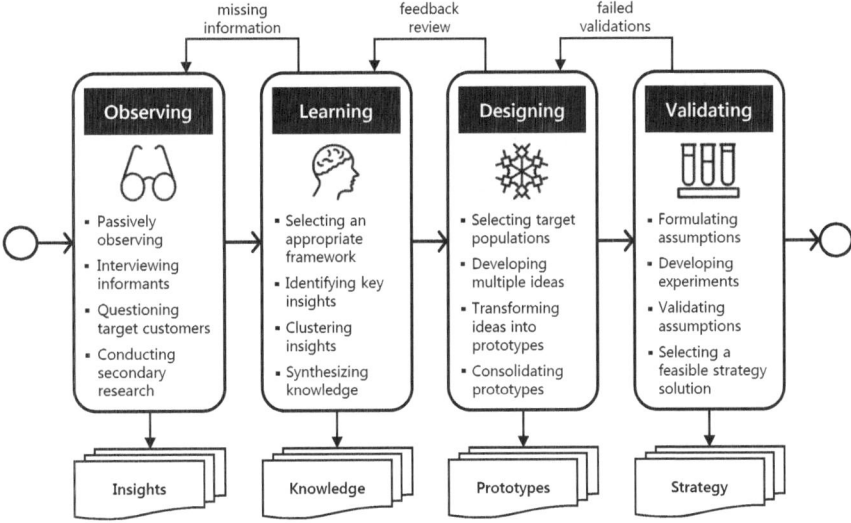

**Fig. 2.3** Alternative representation of the design thinking for strategy or DTS process

adjusting the learning step outcome of the process. The process is iterated until there are no more open questions that could invalidate the selected strategy or its characteristics. DTS proceeds in an agile, just-in-time way, avoiding analysis that do not contribute to the designed strategy.

Figure 2.3 displays a more graphical representation of the DTS focusing on the activities and expected outcomes of each of the four steps.

## References

Arnheim, R. (1969). *Visual thinking*. Berkeley, CA: University of California Press.

Arnold, J. E. (1959). *Creative engineering*. Lecture notes. Stanford, CA: Stanford University.

Brown, T. (2009). *Change by design: How design thinking transforms organizations and inspires innovation*. New York, NY: HarperCollins Publishers.

Buchanan, R. (1992). Wicked problems in design thinking. *Design Issue, 8*(2), 5–21.

Cross, N. (1982). Designerly ways of knowing. *Design Studies, 3*(4), 221–227.

Cross, N. (2006). *Designerly ways of knowing*. London, UK: Springer.

Cross, N. (2011). *Design thinking*. London, UK: Bloomsbury Academy.

Diderich, C. (2018). Understanding the value to design thinking to innovation in banking. *Journal of Financial Transformation, 48*, 64–73.

Eppinger, S. D., & Ulrich, K. T. (1995). *Product design and development*. New York, NY: McGraw Hill.

Koberg, D., & Bagnall, J. (1972). *The universal traveler: A soft-systems guide to creativity, problem-solving, and the process of reaching goals*. San Francisco, CA: William Kaufmann.

Lawson, B. (1980). *How designers think*. Oxford, UK: Butterworth Architecture.

Liedtka, J., Salzman, R., & Azer, D. (2017). *Design thinking for the greater good*. New York, NY: Columbia University Press.

Liedtka, J., & Ogilvie, T. (2011). *Designing for growth*. New York, NY: Columbia University Press.

Martin, R. (2009). *The design of business: Why design thinking is the next competitive advantage*. Boston, MA: Harvard Business Press.

McKim, R. H. (1973). *Experiences in visual thinking*. Pacific Grove, CA: Brooks-Cole Publishing.

Rowe, P. G. (1987). *Design thinking*. Cambridge, MA: MIT Press.

Sawyer, R. K. (2012). *Explaining creativity*. Oxford, UK: Oxford University Press.

Schneider, J., & Stickdorn, M. (2011). *This is service design thinking*. Hoboken, NJ: Wiley.

Simon, H. A. (1968). *The science of the artificial*. Cambridge, MA: MIT Press.

Tschimmel, K. (2012). *Design thinking as an effective toolkit for innovation*. In Proceedings of the XXIII ISPIM Conference: Action for Innovation. Innovation from Experience. Barcelona, Spain.

Vlaskovits, P. (2011). Henry Ford, innovation, and that "Faster Horse" quote. *Harvard Business Review*. https://hbr.org/2011/08/henry-ford-never-said-the-fast.

# Revisiting the Business Model Canvas as a Common Language

3

*Fact is, inventing an innovative business model is often mostly a
matter of serendipity*—Gary Hamel

With the use of design thinking to address business challenges, the concept of business model has gained traction. Although there does not exist a common definition of what a business model is, Shafer et al. (2005) have identified four components that are found in all definitions of a business model. Any business model includes

- a set of *strategic choices*,
- a link between customers and value offered to them, called the *value network*,
- *capabilities and resources* to create value for both the customers and the firm, and
- a *mechanism to capture the created value* and turn it into profits.

The academic literature on business models has been developed mainly in three silos (Zott et al. 2011):

(1)  Focus on leveraging information technology in organizations.
(2)  Manage innovation and technology capabilities.
(3)  Address specific strategic issues.

Despite these different approaches, most business model definitions focus on describing, in a holistic way, how a firm does business and generates profits. They center around the firm and its customers rather than focusing on competitors or the external environment. In contrast with other models for describing how a firm does business, the business model framework takes a customer-centric approach, seeking to explain how value is created and captured for both the customer and the firm. Despite a business model having many similarities with a business strategy, it is not a novel way of describing a firm's strategy. A firm's strategy is broader than its business model. Indeed, a strategy includes competitive positioning and differentiation, on top of the business model characteristics around customers, their

© Springer Nature Switzerland AG 2020
C. Diderich, *Design Thinking for Strategy*, Management for Professionals,
https://doi.org/10.1007/978-3-030-25875-7_3

jobs-to-be-done, as well as capabilities and resources needed to generate profits. A strategy describes reactions to competitive moves and interactions with the surrounding environment. But no strategy can survive and prosper without an underlying sound business model.

A definition of business model that is rooted in addressing strategic issues, rather than focusing on innovation or technology, is needed to support the design thinking for strategy process with a common language. Based on Osterwalder and Pigneur's (2010) business model canvas description, the business model provides a holistic characterization of a firm focusing on four components:

(1) Customers.
(2) Offerings.
(3) Capabilities.
(4) Financials.

The essential details of a firm's value creation and capturing capabilities for its stakeholders are outlined. In contrast to other approaches, the level of detail of the business model is defined by the context in which it is used rather than by the framework itself. This creates flexibility without giving up clarity.

## 3.1  The Role of the Business Model in the Context of Strategy Design

Academic research has successfully studied the concepts behind strategy since the 1960s without the need to have recourse to the notion of business model. But strategy work has remained expert work, making it hard, if not impossible, to integrate multiple stakeholders with diverse backgrounds from various levels of the organization. The business model framework provides relief to that challenge by offering two key added-values:

(1) The business model framework provides a *common language* that allows describing and discussing how a firm creates, captures, and delivers value to all its stakeholders, including customers and well as shareholders. It supports a mutual understanding of the idiosyncrasies of a firm in a holistic way and allows identifying its competitive positioning.
(2) The business model framework provides a *customizable level of abstraction* that can be tailored to the challenge at hand. It supports the top-down design and validation of a strategy ensuring consistency at all layers of the strategy design process.

As such, the business model framework provides a tool, a model, similar in concept to Porter's five forces model (Porter 1979) or the value chain model (Porter 1985), focusing on identifying key properties of a firm's strategy. But it is more pragmatic, broader in scope, and puts a significant focus on customers and their relationship with the firm. It is used in different flavors throughout the strategy design process.

The business model framework I use to support the strategy design process is based on the original work of Osterwalder and Pigneur (2010). It has been adapted based on academic insights from strategy research and practical experience from using the model in day-to-day strategy work. Two levels of abstraction of the business model framework are used to support the strategy design process:

(1) The *lightweight business model*, first introduced in 2015 (Diderich 2017), focuses on identifying the key dimension along which a firm competes. It can be perceived as the operationalized equivalent of the vision and mission statements found in traditional strategies. It allows formulating the firm's goals using concrete statements rather than abstract phrases.

(2) The *detailed business model*, extending the original business model canvas (Osterwalder and Pigneur 2010), supports designing the different elements ensuring that the strategy is desirable, feasible, and viable. It aids concretizing the strategic focus of the firm developed using the lightweight business model.

A major advantage of applying business models as a tool to design strategies over using traditional frameworks, is that they can be used for describing the current situation as well as the future target state in a holistic, concise, and easy to understand way. Rather than using different models to characterize distinct parts of the business model, like the value proposition model (Osterwalder et al. 2014) or the value chain model (Porter 1985), the strategy design process introduced in this book relies on the same framework, at different levels of abstraction, that is zooming in and out of it, depending on the needs throughout the overall strategy design process. This ensures a holistic view of the firm and its strategy.

Although the delineation between design and implementation at the strategy design level is often blurry, business model design and business model implementation are differentiated. Indeed, the goal of the business model design is to support answering questions around "what to do", whereas business model implementation concentrates on the "how to do it" questions. Implementing a business model typically relies on specifying the firm's operating model, that is, operationalizing its strategy.

## 3.2   The Lightweight Business Model

The lightweight business model supports the identification of the high-level characteristics on which a firm's competitive positioning is defined. It focuses on what makes a firm in particular, competitors in general, and an industry as a whole, valuable and unique. As shown in Fig. 3.1, the lightweight business model constitutes of four building blocks, called *components*:

(1) Customers.
(2) Offerings.
(3) Capabilities.
(4) Financials.

The lightweight business model provides the right vocabulary to talk about strategy in ways that help understand what causes it to succeed (Christensen et al. 2016a, 2016b). Each of the four components describes a key characteristic of a firm or an industry. Strategy development starts by selecting exactly one component on which the firm puts its primary competitive focus, the *strategic focus*. The framework has been designed with simplicity in mind, allowing executives and managers, not specifically trained in strategic thinking, to express and structure facts, opinions, and ideas related to their firm, their competitors, and the industry they are operating in. Notwithstanding, the design of the lightweight business model is based on academic rigor.

### 3.2.1   Rationale and Conceptual Details

Consistent with the work of Porter (1985), the lightweight business model distinguishes between competing on differentiation (top part of the lightweight business model, as illustrated in Fig. 3.1) and competing on price (bottom part of the business model, as illustrated in Fig. 3.1).

Horizontally, the lightweight business model distinguishes between customers (and their jobs-to-be-done), offerings (and the associated value propositions), and capabilities (skills as well as resources) as potential dimensions to provide a competitive differentiation. This subdivision is aligned with the work by Treacy and Wiersema (1995) on the strategic value disciplines model, which states that any successful firm must show superior characteristics along one of the three dimensions, customer intimacy, product innovation, or operational excellence, and be proficient in the two other dimensions.

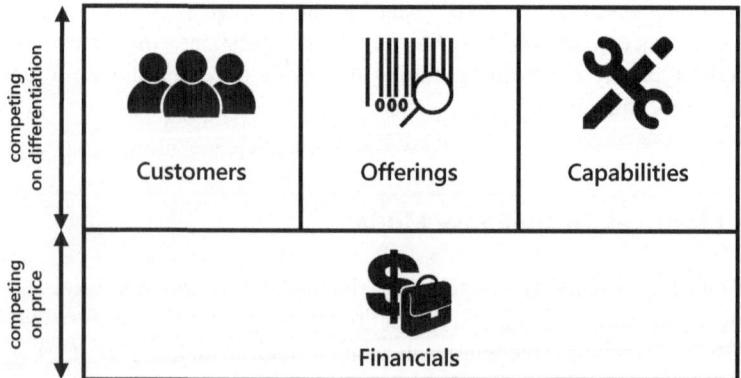

**Fig. 3.1**  The lightweight business model

### 3.2.1.1 Customers

Every firm requires customers. But a firm that tries to serve every customer will fail, as many examples throughout history have shown. The customer component describes which customers to serve and which not to serve. Ideally, personas (Liedtka and Ogilvie 2011) should be used to characterize the targeted customers in a human way. The key jobs-to-be-done (Christensen et al. 2016a) of the identified personas need to be described by taking a customer perspective, rather than a firm perspective. To do so, the customer decision journey (Court et al. 2009) and the journey map (Liedtka et al. 2014) serve as frameworks for structuring insights used to defining the customer component. Firms focusing primarily on customers and their existing needs, felt pains, and thought after gains, are called customer-centric firms.

### 3.2.1.2 Offerings

The offerings component focuses on describing what promises are made to the customers and how satisfying those promises creates value. Firms focusing on differentiation through unique offerings put innovation, sometimes even specific inventions, at the center of their strategy. Innovation does not necessarily have to come from technology. It may also be rooted in visual design, user experience, or even in identifying new customer needs to be satisfied. Excelling along the offerings component requires identifying and addressing needs that customers are yet to become aware of. New needs are created rather than existing needs satisfied.

### 3.2.1.3 Capabilities

The capabilities component describes, based on the value chain, the pivotal qualities of the firm, that is, its capabilities, both in terms of skills and resources, which, if unavailable, will make it fail. Fast-follower strategies are usually based on a capabilities focused strategy (Markides and Geroski 2004). Integrated strategies, for example those that build on supply chain management, qualify for a capabilities based strategic focus. They aim at leveraging internal resources, skills, and expertise, targeting existing customer needs and extending the existing products and services portfolio.

### 3.2.1.4 Financials

The financials component describes the key characteristics underlying how the firm expects to generate revenues and handle costs, from a strategic perspective. Firms aiming at being discounters need to show superior characteristics along the financials component. Especially in the age of platform and so-called freemium strategies, understanding the financials component is key.

---

Diverse types of strategies can directly be related to any of the four components. Successful firms excel in exactly one of the four components. They define their competitive advantages along exactly that one component. Firms that attempt to compete along more than one component often fall in the "stuck in the middle" trap, where being stuck in the middle comes from attempting to compromise, resulting in customers no longer knowing what the firm stands for and what to expect.

**Example** To illustrate the concept of the lightweight business model, consider Apple, and for the sake of simplicity, its mobile devices business, that is phones, tablets, and watches. The goal behind using the lightweight business model is understanding the key traits that make-up the foundation of Apple's strategy, understanding how Apple aims at creating value for its customers, its shareholders, and understanding how Apple differentiates itself from competitors, like Samsung or Huawei.

First, have a look at the customers targeted by Apple, wealthy private clients seeking to be always online and valuing design as well as usability over technology. In addition, Apple customers seek status and want to stand apart from the mass. The lightweight business model supports conciseness in describing of these insights. To satisfy diverse customer needs, Apple offers mobile devices in different sizes, from a 1"5-watch to a 12"9-inch tablet. The focus is primarily on brand, design, and functionalities rather than on the latest technological innovations. In addition, the offerings give a feeling of quality, rather than value for money. When looking at the third component, capabilities, Apple exhibits a unique capability of identifying and implementing new trends. Apple is skilled in identifying new unmet needs and satisfying them with well-designed and implemented technologies. In contrast to competitors, as the lightweight business model shows, Apple is not competing on technology innovation, but on user experience innovation. Finally, Apple's financial success is due to it being able to charge a relatively high price tailored to specific features selected by the customer and related to the perceived rational and especially emotional value of the offerings rather than relying on a cost-plus-margin pricing model. It separates value-added production activities from commodity activities that can cheaply be outsourced. Figure 3.2 summarizes Apple's lightweight business model.

| Customers | Offerings | Capabilities |
|---|---|---|
| • Wealthy retail customers worldwide valuing design <br> • Customers seeking status and wanting to stand apart from the masses <br> • Customers valuing usability and technology | • Digital watches <br> • Mobile phones <br> • Tablets <br> • Recognized brand, design, and functionalities <br> • Focus on high quality | • Brand development <br> • Trend identification <br> • Identifying unmet customer needs <br> • Product design <br> • App platform ecosystem |

| Financials |
|---|
| • Fixed price for device based on specific features selected by the customers <br> • Different prices in different countries to exploit local purchasing power <br> • App platform fee <br> • Retrocessions from partners, especially mobile phone operators |

**Fig. 3.2** Lightweight business model describing Apple's mobile devices business unit

Once the lightweight business model is defined, it is easy to structure the discussion and design around, for example, how a firm could compete in a market with a strong Apple presence. The question boils down to in which of the four components of the lightweight business model the firm wants to compete against Apple, either by being different or by being superior. Should the firm, for example, target a different and large enough customer segment to which the firm has access and a better understanding of their jobs-to-be-done than Apple. Or should the firm compete on the offerings component by differentiating through providing unique, state-of-the-art technology features, rather than following Apple's user experience and visual design path. Or should the focus be on a low-cost mobile device strategy as viable strategic alternative?

As illustrated in this short example, the lightweight business model allows structuring the discussion around a firm's strategic focus in a very effective way.

## 3.3   The Detailed Business Model

Even after the lightweight business model of a firm has been defined, there remain numerous open questions. The *detailed business model*, an extended version of the lightweight business model, based on the business model and value proposition canvases (Osterwalder and Pigneur 2010; Osterwalder et al. 2014), is used to successfully identify and address these questions.

The detailed business model retains the structure of the lightweight business model, distinguishing between customers, offerings, capabilities, and financials and deepens them. Figure 3.3 illustrates the overall structure of the detailed business model.

In contrast with the original business model canvas from Osterwalder and Pigneur (Osterwalder and Pigneur 2010), the detailed business model is designed to be read, designed, and validated, from left to right, starting with the customers. This is consistent with the western society, left to right, reading and writing habit. Ideally different colors are used for different business model components—green for customers, red for offerings, blue for capabilities, and yellow for financials.

To describe an existing or design a new business model supporting a strategy, it is recommended to start with defining the elements of the component along which the firm exhibits superiority and its competitive advantage, that is, its strategic focus. For example, in the Apple example, the description best starts with the value proposition and offerings, moving to capabilities, followed by customer related aspects, finishing with supporting costs and revenue streams.

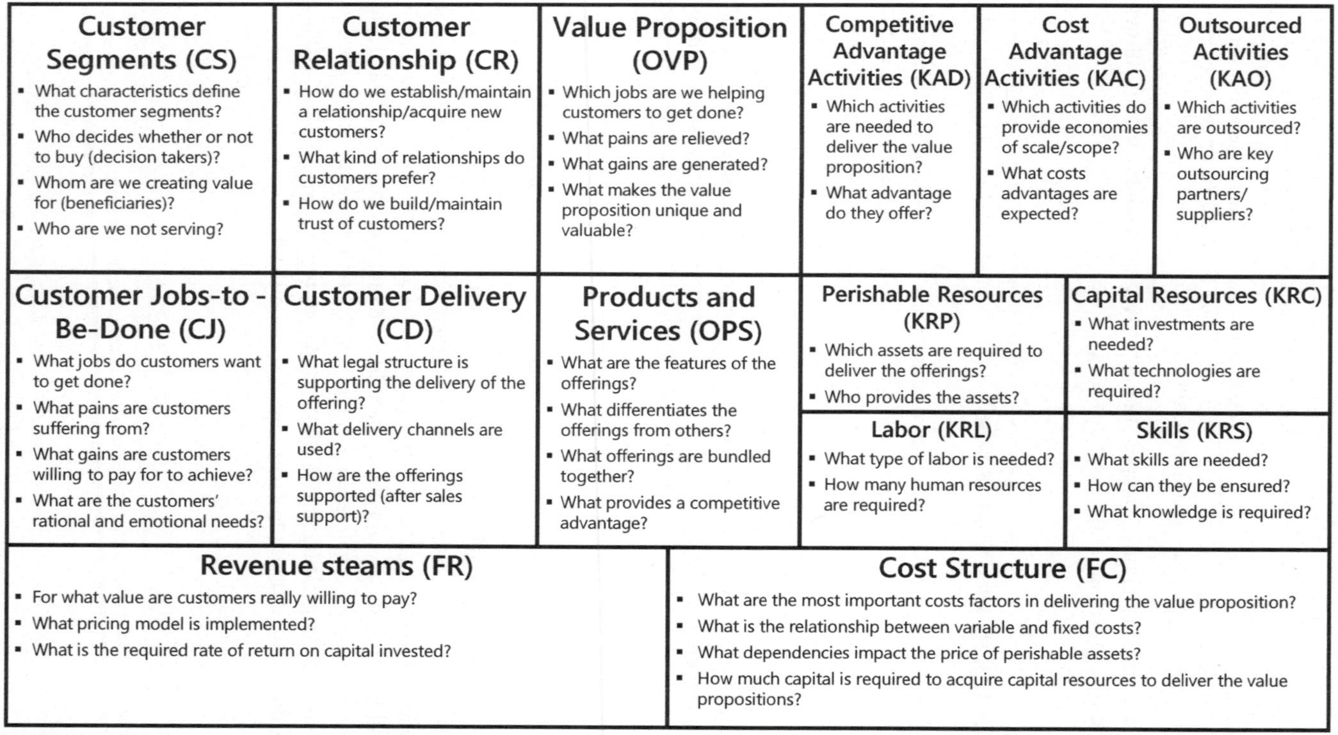

**Customer Segments (CS)**
- What characteristics define the customer segments?
- Who decides whether or not to buy (decision takers)?
- Whom are we creating value for (beneficiaries)?
- Who are we not serving?

**Customer Relationship (CR)**
- How do we establish/maintain a relationship/acquire new customers?
- What kind of relationships do customers prefer?
- How do we build/maintain trust of customers?

**Value Proposition (OVP)**
- Which jobs are we helping customers to get done?
- What pains are relieved?
- What gains are generated?
- What makes the value proposition unique and valuable?

**Competitive Advantage Activities (KAD)**
- Which activities are needed to deliver the value proposition?
- What advantage do they offer?

**Cost Advantage Activities (KAC)**
- Which activities do provide economies of scale/scope?
- What costs advantages are expected?

**Outsourced Activities (KAO)**
- Which activities are outsourced?
- Who are key outsourcing partners/suppliers?

**Capital Resources (KRC)**
- What investments are needed?
- What technologies are required?

**Customer Jobs-to-Be-Done (CJ)**
- What jobs do customers want to get done?
- What pains are customers suffering from?
- What gains are customers willing to pay for to achieve?
- What are the customers' rational and emotional needs?

**Customer Delivery (CD)**
- What legal structure is supporting the delivery of the offering?
- What delivery channels are used?
- How are the offerings supported (after sales support)?

**Products and Services (OPS)**
- What are the features of the offerings?
- What differentiates the offerings from others?
- What offerings are bundled together?
- What provides a competitive advantage?

**Perishable Resources (KRP)**
- Which assets are required to deliver the offerings?
- Who provides the assets?

**Labor (KRL)**
- What type of labor is needed?
- How many human resources are required?

**Skills (KRS)**
- What skills are needed?
- How can they be ensured?
- What knowledge is required?

**Revenue steams (FR)**
- For what value are customers really willing to pay?
- What pricing model is implemented?
- What is the required rate of return on capital invested?

**Cost Structure (FC)**
- What are the most important costs factors in delivering the value proposition?
- What is the relationship between variable and fixed costs?
- What dependencies impact the price of perishable assets?
- How much capital is required to acquire capital resources to deliver the value propositions?

**Fig. 3.3** Detailed business model

### 3.3.1   Rationale and Conceptual Details

The detailed business model is made up of 15 sub-components (4 related to customers, 2 related to offerings, 7 related to capabilities, and 2 related to financials), called *elements*, of the four components from the lightweight business model. Each element focuses on a specific trait of the firm and takes a unique perspective, either internal or external.

#### 3.3.1.1   Customers
The customer component from the lightweight business model is refined into four elements, each focusing on a specific aspect of the customer relationship.

*Customer Segments (CS).* The customer segments element defines groups of customers that have similar needs and can be served in a comparable way. Customer segments should be defined around common jobs-to-be-done (Christensen et al. 2016a) using personas (Liedtka and Ogilvie 2011), rather than relying on demographic properties. Defining a given customer segment does not necessarily mean that this segment will be served by the firm through an offering. Customer segmentation needs to consider who may potentially use and benefit from the offerings, who decides about a purchase, and who actually pays for it, that is, writes the check.

Key in successfully defining customer segments is ensuring homogeneity and size, two attributes that are often antagonisms. In addition, successfully defining customer segments is tightly related to the other three customer elements. Identifying the most appropriate customer segments is therefore an iterative process. Customer segments should be defined independently of offerings and value propositions to avoid too narrow definitions and missing out on opportunities, especially in a customers-focused strategy.

> **Example** Sales executive John travels 80% of his time and needs to be able to access his firm's CRM system anytime and anywhere, in addition to always being reachable during office hours. This is a typical description of a persona defining one customer segment for a mobile phone operator. Another would be Daisy, an outspoken millennial who wants to be up-to-date with the latest gossip and likes to talk long hours to her friends on the phone.

*Customer Jobs-to-Be-Done (CJ).* The concept of customer jobs-to-be-done is a key concept of the strategy design process. It has been introduced by Christensen et al. (2016a, 2016b) and focuses on describing what customers want to achieve. Research has shown that customers buy perceived value that helps them satisfy their jobs-to-be-done rather than buy outright products or services.

> **Example** A customer wants to satisfy his thirst, or he wants to kill time, two completely different jobs-to-be-done that could be satisfied by the same offering, a drink, but with two completely different value propositions.

It is key to take the perspective of the customer and not the one of the firm, when identifying jobs-to-be-done. A rational and/or emotional value should be associated with each identified job-to-be-done. As for customer segments, just because a job has been identified, this does not necessarily mean that an associated offering must be provided. Strategy is about choice. Choice requires options to choose from being identified. More often than not, are jobs-to-be-done observed before customer segments are defined. This is consistent with the design thinking philosophy to start by observing (jobs-to-be-done) before learning (customer segments).

*Customer Relationship (CR).* Many strategies fail because they have not taken care of the customer relationship element. Customers are of value only if they can be reached, acquired, and retained. The customer relationship element identifies how a customer relationship is established and maintained. In many business models, branding plays an important role in creating and sustaining a customer relationship. Again, as for all customer elements, the customer relationship element must be addressed from a customer perspective. What relationship do customers value? Is it a walk-in relationship, is it based on network effect, or is it based on a brand, to name just a few possible customer relationship approaches?

More important than building new customer relationships is keeping existing relationships. The customer relationship element must define how the relationship is nurtured over time. Estimates show that it is, on average, between four to six times costlier to gain a new customer than to keep an existing one. In addition, the probability of selling to existing customers is 14 times higher than to selling to new customers (Bendle et al. 2016).

> **Example** Nespresso nurtures the relationship with its customers through its unique online club. From a customer perspective, value is generated and delivered by making available unique offers through the club relationship infrastructure not available elsewhere.

*Customer Delivery (CD).* Once a customer has been acquired, their jobs-to-be-done identified, and an offering sold, it must be delivered. The customer delivery element details the key characteristics of these activities. Regulatory and legal requirements must be considered. For example, consumer goods need specific packaging material not to be used. In the fund management industry for example, the proper legal structure, based on the customer segments, must be chosen. The actual delivery process, the channels used, must be aligned with the customer needs.

> **Example** When home delivering valuable goods, customers need to be at home or an alternative secure delivery approach must be provided. Delivering the goods in a tamper proof box securely attached with a chain and a code to the customer's mailbox may be a solution.

When describing the customer delivery element, after-sales services need to be considered. Especially when something goes wrong, or customers have questions or complaints, they need to be able to reach the firm and get their matter taken care of. As satisfied customers tend to become recurring customers and/or give referrals, it is important to offer a great delivery experience. You never get a second chance to make a first impression, is an often-cited quote by Will Rogers, applicable in this context. It should be captured by the definition of the customer delivery element of the detailed business model.

### 3.3.1.2 Offerings

Many inexperienced strategists only focus on the offerings component, that is, the products and services sold. There is much more to a business model than any specific product or service. The offerings component is made up of two elements, providing two complementary perspectives of the same insights.

*Value Proposition (OVP).* The value proposition describes the offerings from a customer perspective, focusing on the jobs-to-be-done satisfied and the value, both rational and emotional, provided. The value proposition element is tightly linked to the customer journey of the served customer segments. Many strategists make the mistake to look at the value proposition from the firm and their own personal perspective and forget the customer in the equation. This often leads to features no customer wants and/or is willing to pay for. Note that customers pay for perceived value and not for goods or services, even if they say so. This is even the case in commodity goods markets.

*Products and Services (OPS).* The products and services element describes the rational characteristics and features of the offerings, as produced by the firm and delivered though the customer delivery element to the targeted customer segments. It also describes possible bundling. A specific focus should be put on those characteristics that makes the products and services different or superior to similar products and services from competitors.

### 3.3.1.3 Capabilities

The capabilities component of the detailed business model focuses on the internals a firm needs to implement to provide the products and services to customers and deliver the value propositions promised. It is the part of the detailed business model that is closest related to traditional strategy design, especially resources-based views strategies (Barney 1991). The capabilities component can be subdivided into two sub-components, that is,

- those elements that describe the *activities* to be performed to produce and deliver the offerings to customers and position the firm relative to its competitors, and
- those elements that describe the required *resources* to perform these activities.

*Key Activities (KA).* Based on Porter's (1985) theory of competitive advantage and associated value chain arguments, key activities can be classified into three categories:

(1) *Activities providing a Differentiation Advantage (KAD)*, either by performing or combining activities in a different way from those of competitors or by performing the same activities in a superior way through a better or different exploitation of the underlying resources.

(2) *Activities providing a Cost Advantage (KAC)*, through either exploiting economies of scale and/or scope or using resource and skills more efficiently and/or effectively than competitors. Sometimes the cost advantage stems from managing the interface with an outsourcing provider. This is often the case for example, for payroll handling. In other cases, cost advantages may be of regulatory nature.

(3) *Outsourced Activities (KAO)*, that neither provide a differentiation, nor a cost advantage, but are essential to produce and deliver the offering.

In an ideal world, any activity falls into one of those three categories. In practice, nevertheless, there exist activities that a firm may decide to perform itself, although they could be outsourced.

*Key Resources (KR).* Based on Drucker (2006), four distinct types of key resources can be identified:

(1) *Perishable Resources (KRP)*, that is, assets that are consumed during the production and delivery of the offerings. In a digital business model, perishable assets may be content, like news. For an airline, fuel, as well as served food are key perishable resources. An important property of key perishable resources is, that they are usually provided by external suppliers and, as such, are not under the direct control of the firm. During strategy design, possible substitutes for key perishable resources need to be identified and their potential shortage impact on the sustainability of the strategy addressed. The pricing power of suppliers needs to be considered as argued by Porter (1979) in his five forces model. Perishable resources may also be needed to manage the customer relationship, like marketing goodies, or for delivering the offerings, like postal services.

(2) *Capital Resources (KRC)* result from investments in technology, infrastructure, or equipment. Capital resources may be intangible, for example, a brand or intellectual property. Capital resource support the activities to produce and deliver the offerings. For Google, their search algorithm is a key capital resource. The value of capital resources may or may not depreciate over time and their potential replacement and values need to be considered in the overall strategy design.

(3) *Labor Resources (KRL)*, define the human resources required to produce and deliver the offerings. They are complementary to capital resources. A key indicator of labor resources is quantity. Robots, or machines, assuming they do not themselves provide a differentiating element, can be qualified as labor, rather than capital. During strategy development, strategy designers need to decide whether to realize an activity through labor resources or capital resources. Answering strategic questions around automation often relate to what type of resources are needed.

(4) *Skill Resources (KRS)*, in contrast to labor resources, focus on knowledge rather than quantity of workforce. The skill resources element describes the skills and knowledge that are required to execute the activities necessary to produce and deliver the offerings. Skill resources, in contrast with labor resources, are usually scarce. It is important to define how the skills are acquired and retained over time. The focus must be on those skills that support the strategy's differentiating activities.

### 3.3.1.4 Financials

The last but not least important component of the detailed business model takes care of the financial aspects of the firm, that is, revenues and costs.

*Revenue Streams (FR)*. There are two aspects that subsume the revenues element, that is,

- the revenue model, and
- prices and volume over time.

In traditional strategy development, for example, when relying on a DuPont analysis,[1] only the price and volume over time aspects are considered. In modern strategy design processes, developing innovative pricing models becomes increasingly relevant. A unique pricing model may provide a competitive advantage. More important, pricing models must support the jobs-to-be-done way of thinking and tie the customers to the firm. Indeed, pricing can be looked at from a cost center or from a profit center perspective. The customers' background and mindset have an influence on selecting the most appropriate pricing model.

**Example** Artisans typically prefer pricing models related to a unit of size, while lawyers think in units of time, and privateers often prefer an all-in single figure pricing model.

---

[1]The DuPont analysis is an expression which breaks ROE (return on equity) into its parts. Its name comes from the DuPont Corporation that started using this formula in the 1920s. DuPont explosives salesman Donaldson Brown invented this formula in an internal efficiency report in 1912.

Recently, recurring pricing models have become in vogue as they support the customer relationships element. Other pricing models, like those implemented by private TV stations differentiate between the customer whose jobs-to-be-done are addressed, the viewers, and those who are writing the check, the advertisers. Investment firms implement pricing models that relate fees to investment performance, at least in part.

*Cost Structure (FC).* On the cost side, expenses can be subdivided into those used to pay for perishable assets and labor, and those used to pay for investments, needed to support the capital resources and the acquisition of skill resources. The main difference between the two categories of expenses are the timing of their cash flows. Financial engineering can be applied to adjust the timings of cash flows, like found in sale and leaseback contracts. Costs need to support all capabilities required and must allow, in combination with revenues, to generate profits that are no smaller than the cost of capital, defined by the inherent risk of the resulting strategy.

### 3.3.2   Relations Between Elements of the Detailed Business Model

There must exist a one-to-many, or at least a one-to-one relationship between the products and services and value proposition elements. Each value proposition must relate to one or more jobs-to-be-done associated with one or more customer segments, that the firm aims at serving. Similarly, each product and service must be related to activities, assets, and resources. As a firm should not serve everyone or address all jobs-to-be-done, some of the customer segments and/or jobs-to-be-done may not have a relationship to a value proposition. Figure 3.4 illustrates the

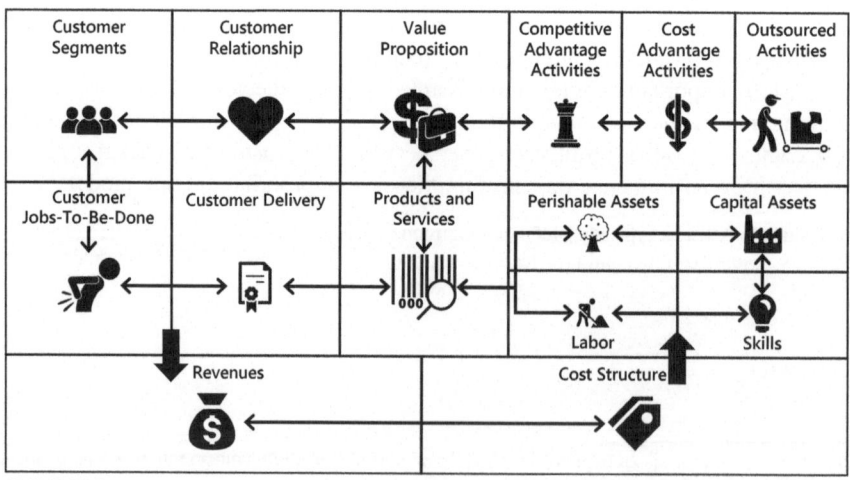

**Fig. 3.4** Link between the different elements of the detailed business model

| Customer Segments (CS) | Customer Relationship (CR) | Value Proposition (OVP) | Competitive Advantage Activities (KAC) | Cost Advantage Activities (KAC) | Outsourced Activities (KAO) |
|---|---|---|---|---|---|
| ▪ Private customers worldwide<br>▪ Wealthy customers not buying their mobile phone on price<br>▪ Customers seeking status and wanting to stand apart from the masses<br>▪ Customers valuing usability and technology | ▪ Apple brand<br>▪ Live streamed and broadly publicized launch events<br>▪ Apple stores<br>▪ Self-service on apple.com<br>▪ iTunes platform for apps | ▪ High quality product design<br>▪ Products implementing new trends<br>▪ Status through brand<br>▪ Convenience / Usability<br>▪ Large set of apps available, including from Microsoft | ▪ Identifying trends<br>▪ Brand development<br>▪ Product design<br>▪ iOS development<br>▪ Executing launch events | ▪ Providing app distribution platform<br>▪ Brand power in attracting partners and contributors to support mobile devices | ▪ Manufacturing of individual parts<br>▪ Device assembly<br>▪ Delivery to customers<br>▪ App development |

| Customer Jobs-to-Be-Done (CJ) | Customer Delivery (CD) | Products and Services (OPS) | Perishable Resources | Capital Resources (KRC) | |
|---|---|---|---|---|---|
| ▪ Looking for differentiating through design and brand<br>▪ Valuing perceived product quality<br>▪ Seeking to be always online<br>▪ Wanting to be able to satisfy a diverse set of needs on a single mobile platform | ▪ Online sales on apple.com<br>▪ Sales through own and third-party stores<br>▪ Cooperation with mobile phone operators<br>▪ Personal support in Apple store – Genius Bar | ▪ Digital watches<br>▪ Mobile phones<br>▪ Tablets<br>▪ Seamless handover between devices | ▪ Mobile phone parts | ▪ Apple brand<br>▪ Intellectual property<br>▪ Product design teams and infrastructure | |
| | | | Labor (KRL) | Skills (KRS) | |
| | | | ▪ Store and genius bar staff | ▪ Branding and marketing<br>▪ Trend spotting<br>▪ Hardware design and software development | |

**Revenue streams (FR)**

▪ Fixed price for devices, tailored to specific features selected by the customer
▪ Different prices in different countries to explicit local purchasing power
▪ Apple care support package
▪ Fees from sales of apps through iTunes platform
▪ Retrocessions from partners, especially mobile phone operators

**Cost Structure (FC)**

▪ Salaries and bonuses of skill providing people
▪ Branding and marketing
▪ Store operations, including genius bar
▪ Paying outsourcing partners and perishable resources

**Fig. 3.5** Detailed business model of Apple's mobile device line of business from an external perspective

relationships between the different elements of the detailed business model. Missing any of those relationships may lead to a failing strategy.

A value proposition can only be characterized as such if

- it addresses one or more customer jobs-to-be-done, rational and emotional needs, perceived pain points, or sought-after gains, and
- the customers are willing to pay for getting the jobs done, the needs satisfied, the pains relieved, or the sought-after gains achieved.

This can be especially challenging if the customers whose jobs-to-be-done are addressed is not the same customer that is paying for getting the job done.

**Example** A wireless pulse oximetry device addresses a nurse's job-to-be-done of avoiding wire cluttering in hospitals. This is a sound offering from the nurse's perspective, but hospital administrators only see costs, and no value, and as such do not identify the offering addressing a job-to-be-done they are willing to pay for.

To illustrate how the detailed business model is used to describe a firm, consider Apple's mobile devices line of business. Figure 3.5 describes the associated detailed business model, as perceived from outside. Note the simplicity and focus used in the description. Only insights that matter and are specific to Apple are described. Information, like, for example, the battery resource is not included in the detailed business model description, as it is not key to Apple's strategy, even though mobile devices don't work without them.

# References

Barney, J. B. (1991). Firm resources and sustained competitive advantage. *Journal of Management, 17*(1), 99–120.

Bendle, N. T., Farris, P. W., Pfeifer, P. E., & Reibstein, D. J. (2016). *Marketing metrics: The manager's guide to measuring marketing performance* (3rd ed.). Upper Saddle River, NJ: Pearson Education.

Christensen, C. M., Hall, T., Dillon, K., & Duncan, D. S. (2016a). Know your customers' "jobs to be done". *Harvard Business Review, 94*(9), 54–62.

Christensen, C. M., Hall, T., Dillon, K., & Duncan, D. S. (2016b). *Competing against luck: The story of innovation and customer choice.* New York, NY: HarperCollins Publishers.

Court, D., Elzinga, D., Mulder, S., & Vetvik, O. J. (2009). The consumer decision journey. *McKinsey Quarterly* (3).

Diderich, C. (2017). Initiating the strategy process using design thinking. *Change Management Strategy eJournal* 9(8). SSRN: https://ssrn.com/abstract=2927941 or http://dx.doi.org/10.2139/ssrn.2927941.

Drucker, P. F. (2006). *The effective executive: The definitive guide to getting the right things done.* New York, NY: Harper Collins.

Liedtka, J., & Ogilvie, T. (2011). *Designing for growth.* New York, NY: Columbia University Press.

Liedtka, J., Ogilvie, T., & Brozenske, R. (2014). *The design for growth field book.* New York, NY: Columbia University Press.

Markides, C., & Geroski, P. A. (2004). Racing to be second. *Business Strategy Review, 15*(4), 25–31.

Osterwalder, A., & Pigneur, Y. (2010). *Business model generation.* Hoboken, NJ: Wiley.

Osterwalder, A., Pigneur, Y., Bernarda, G., & Smith, A. (2014). *Value propositon design*. Hoboken, NJ: Wiley.

Porter, M. E. (1979). How competitive forces shape strategy. *Harvard Business Review, 57*(2), 137–145.

Porter, M. E. (1985). *Competitive advantage*. New York, NY: The Free Press.

Shafer, S. M., Smith, H. J., & Linder, J. C. (2005). The power of business models. *Business Horizons, 48,* 199–207.

Treacy, M., & Wiersema, F. (1995). *The discipline of market leaders: Choose your customers, narrow your focus, dominate your market*. New York, NY: Perseus Books.

Zott, C., Amit, R., & Massa, L. (2011). The business model: Recent developments and future research. *Journal of Management, 37*(4), 1019–1042.

# Part II
# A Structured Approach to Strategy Development

# Gaining a Collective Understanding of the Strategy Development Challenge

**4**

*It always seems impossible until it is done*—Nelson Mandela

In Lewis Carrolls' novel *Alice's Adventures in Wonderland* (Carroll 1865), Alice asks the cat which way to go? To which the cat replied, "that depends on where you want to go?" Similarly, in strategy design, it is important to have a goal in mind before starting. This goal, broadly speaking, is made up of two characteristics, that is,

- the *target industry* in which to compete, and
- *guiding principles* to follow.

When talking about industry, it is important to consider the industry as seen from the customers and their jobs-to-be-done perspective. This is especially relevant when aiming at extending or even disrupting the core industry a firm is competing in. This means, for example, focusing on transportation rather than automobiles, overnight stays rather than hotels, or emergency services rather than hospitals, to name just a few. Defining the industry is at the heart of defining the scope of any strategy design activity. In some cases, it may be appropriate to aim at a technology or a customer segment, rather than an industry as target, if the focus is on inventing something new.

Defining the two characteristics, target industry and guiding principles, sets the stage for designing a firm's strategy. Combined with identifying the key stakeholders that need to be involved at one point or the other during the strategy design process, an initial budget, an expected timeline, an innovation culture, inherent risks, as well as an assessment of the capacity to change of the firm, they form the *strategy brief*. The strategy brief is a short document, prepared by executing process B, focusing on ensuring that everyone starts on the same page. It is written by the strategy team members and confirmed by the stakeholders responsible for the firm's strategy.

Although usually a static document, the strategy brief may be updated during the strategy design process if the findings warrant it. In such a situation it needs to be re-confirmed by the stakeholders responsible for the strategy and communicated to all other stakeholders involved, especially the strategy design team.

© Springer Nature Switzerland AG 2020
C. Diderich, *Design Thinking for Strategy*, Management for Professionals,
https://doi.org/10.1007/978-3-030-25875-7_4

**Process B—Strategy Brief**

B.1  Defining the strategy project set-up:

(1)  Identifying key stakeholders and their roles, including strategy project team members
(2)  Fostering an innovation culture
(3)  Defining an initial high-level budget and an expected timeline
(4)  Assessing the firm's capacity for change as well as determining potential risks arising from strategic decisions

B.2  Identifying the target industry in which the firm aims at competing
B.3  Documenting the guiding principles to adhere to

## 4.1  Strategy Project Set-up

During the strategy project set-up, the environment in which the new or revised strategy is designed, agreed upon, and communicated in a trusted and transparent way is defined.

### 4.1.1  Identifying Key Stakeholders and Their Roles

It is best practice to start by identifying all stakeholders involved in the strategy design process and their expected roles at the forefront in a stakeholder map. Stakeholders actively involved in developing the strategy can be classified into four categories:

(1)  *Decision takers*, responsible for approving the outcome of the different layers of the strategy design process, the milestones.
(2)  *Strategy designers*, responsible for designing the target strategy, focusing on content and its validation.
(3)  *Experts* or *interpreters*, coming from diverse backgrounds and providing fresh ideas from different industries.
(4)  *Process supporters*, managing and supporting the strategy design process.

During the execution of the strategy design process, multiple additional stakeholders are involved, including customer, suppliers, and regulators. It is not necessary to identify all stakeholders at the strategy brief level of the strategy design project. The *stakeholder map*, sometimes called stakeholder list, is an organic document that grows throughout the strategy project.

## Tool—Stakeholder Map

The stakeholder map is a document, part of the strategy brief, describing all stakeholders involved in the strategy design process and the formal, as well as informal, relationships between them. It determines, for each stakeholder,

- their *role* within the strategy design process,
- their *relevance* to success and power to influence success, and
- their *stands* with respect to change.

Figure 4.1 illustrates the typical structure of a stakeholder map. Closest to the center are those stakeholders that are key to success of the strategy design process, surround by most relevant input providers. The outer circle is made-up of stakeholders that are involved at one point or another during the strategy design process, rather than on a continuous base. Nevertheless, they are important. Failing to consider them may, as has been the case many times in the past, derail a strategy design project that was considered sound.

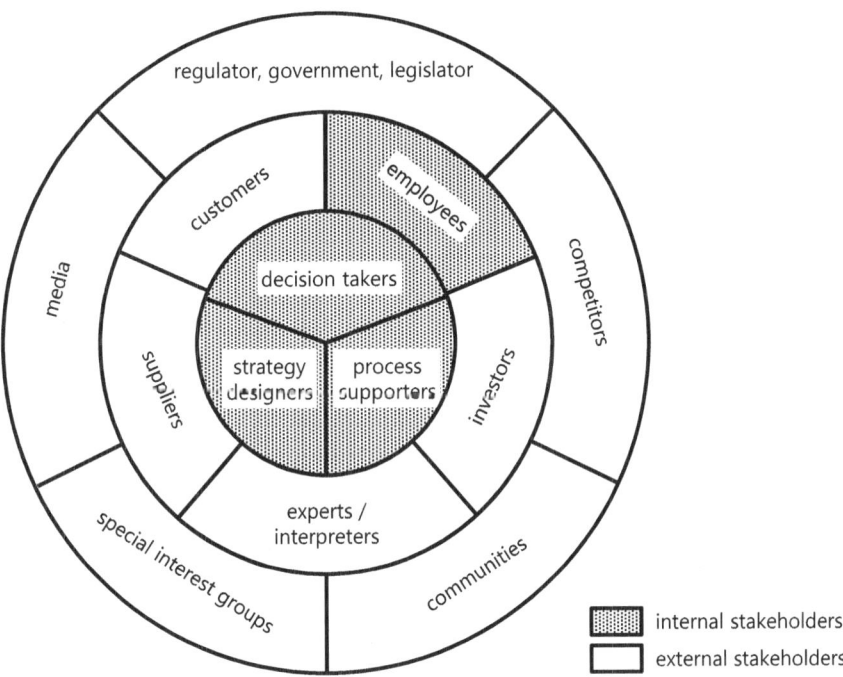

**Fig. 4.1**  Typical structure of a stakeholder map

### 4.1.1.1  Decision Takers

Depending on the applicable national legislation and the firm's governance structure, *decision takers* are members of the board of directors, the management board, or the executive committee. Strategy related decisions must never be delegated to a lower level in the hierarchy of the firm.

To be successful, it is critical that the most important decision takers are involved, or at least informed, all the way along the strategy design process. Decisions, especially at milestones, should always be taken by the same individuals or teams. This guarantees consistency and minimizes the risk of derailment due to so-called politics. Relying on external experts or consultants for decision making must be avoided. Only decisions taken by convinced decision takers having gained the required insights are able to support the decisions made, leading to sustainable success. The group of decision takers should be small and proper accountability must be ensured. Voting based decision taking should be avoided, as it leads to mediocrity.

### 4.1.1.2  Strategy Designers

Strategy designers form a small team, called the *strategy team*, composed of five to ten mostly senior strategists, ideally complemented by board members and senior executives. Creative, forward thinking are key skills that strategy team members must exhibit. *Strategy designers* are in charge of designing and validating the strategy. Ideally, all decision takers should be strategy designers, ensuring buy-in into the final outcome. The more the decision takers are involved in the day to day aspects of the strategy design process, the better its results (assuming a constructive mentality). Strategy designers actively take responsibility for the content produced during the various steps of the strategy design process. Key strategy design work must never be outsourced to external experts or consultants.

### 4.1.1.3  Experts or Interpreters

Experts, called interpreters by Verganti (2009), should have diverse backgrounds and provide fresh ideas. They should be people who look and think beyond the obvious. They should come from different industries, but be proficient in similar contexts, acting as bridge-builders. Their role is to offer new insights and distinct perspectives during observing and learning from target populations. They stand for supporting an extreme discourse during ideation. To avoid a selection bias, that is, choosing those experts whose opinion matches one's own, interpreters should be selected before starting the strategy design process. Experts are usually called-upon during specific phases of the design thinking process when their input is most relevant. Often experts are members of the strategy team, although this is not strictly necessary.

#### 4.1.1.4 Process Supporters

The fourth category of core stakeholders, the *process supporters*, is not less important because they discharge the strategy team, but also because they require distinct capabilities, notable process and information structuring skills. The two types of process supporters are:

(1) *Process moderators, facilitators, and coaches*, supporting and moderating the strategy design process, driving the production of the content throughout the strategy design process.
(2) *Process managers*, managing the process, including documentation and communication. They are usually senior employees of the firm, driving its form, especially the timeline and budget of the process. They are responsible for managing the interface between the strategy design process and stakeholders.

Ideally, the strategy design process is facilitated by one or two independent external strategy coaches. These coaches need to be familiar with the strategy design process and the goals to be achieved at each of its layers and steps. They should also feel at ease with the target industry to ensure the right questions are asked but need not be industry experts. The independence of the coaches ensures that the strategy design process is run as objectively as possible, avoiding any biases, from the "not invented here" syndrome, through leaping, fixating, and overthinking, to satisficing, downgrading, and self-censoring (May 2016).

### 4.1.2 Fostering an Innovation Culture

Creativity in strategy design requires an *innovation culture*. The innovation culture should be designed into the strategy brief, rather than being developed during the strategy design process. An innovation culture has not to be mistaken for ping-pong tables, lounges to chill, or free food. It is about recognizing and valuing uncertainty, ambiguity, and allowing for temporary failure. Successful innovation cultures embrace experimentation. They require strategy team members to bring six key qualities to the table (Mootee 2013), that is,

- *intelligence*,
- broad *knowledge*,
- an *open-minded* thinking style,
- a *team player* personality,
- *motivation*, and
- *comfort* in a changing environment.

An innovation culture must not only focus on individuals, but also on the firm as a whole. To be successful, firms aiming at exhibiting a successful innovation culture must address tree key challenges (Govindarajan and Trimble 2005):

(1) The *forgetting challenge*—Innovative firms allow things to be done differently than they were done in the past. This requires overcoming sources of organizational memory, which in many organizations are very powerful, as firms naturally cling towards operating the way they always have done.

(2) The *learning challenge*—Strategy is, by definition, based on facing the unknown. The best way to face this unknown is through experimenting and learning from the outcome of the experiments. Innovative firms excel in the art and science of experimenting and learning from their results.

(3) The *borrowing challenge*—Most firms do not operate on a greenfield. They have access to exiting assets and capabilities. Innovative firms are able to leverage these values without reverting to the existing course of action.

### 4.1.3 Budget and Timeline

Preparing a reliable budget and timeline for strategy development, especially when aiming at a disruptive strategy, is a challenge. Let alone when using an abductive approach such as design thinking, that aims at optimizing resources used in a just-in-time way. This means, that traditional approaches based on formulating business cases and calculating net present values, will fail.

There exist three guiding principles to follow when deciding on an initial budget (internal, as well as external, resources and funds) and a preliminary timeline.

(1) The budget and timeline determinations should focus on the next decision to be made by the decision takers, at the milestone, or even at the process step rather than on the full strategy design process.

(2) Key indicators relevant to supporting the targeted decisions, that is, the strategy brief, the strategic focus (outcome of the foundation layer), the detailed business model (outcome of the business model layer), the competitive advantage, and the to be communicated strategic message (outcome of the competition layer) should be used to derive initial budget requirements and timeline estimates in terms of

- *internal resources* required,
- *external expertise* and manpower needed as well as their expected costs,
- *funds* required to buy data and insights, and
- *scheduling* of the expected activities on the timeline based on resources availability.

(3) Estimates should be refined after each decision step for subsequent steps maintaining the trust of the decision takers.

A separate budgeting and timeline determination process should be conducted for each of the three layers of the strategy design process (see Chap. 5 for further

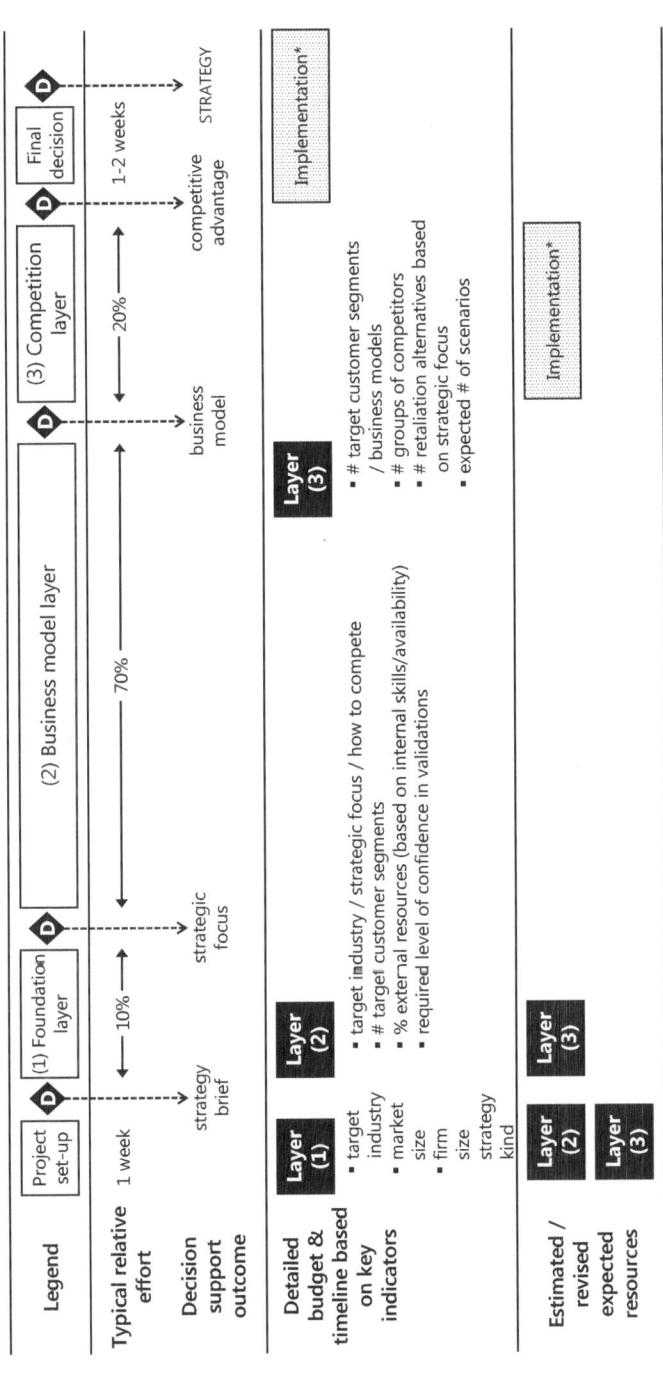

**Fig. 4.2** Typical budget and timeline determination process based on the three-layer strategy design process described in Chap. 5

details on the three-layer strategy design process), starting the next layer budgeting and timelining only at the end of the previous layer as shown in Fig. 4.2. Subsequent budgets and timelines may be decided for each of the four design thinking process steps, observing, learning, designing, and validating, of the business model layer. Sometime, especially in larger strategy projects, it may be sound to manage budget and timeline at the target customer segments and jobs-to-be-done level during the observing and learning step, at the prototype level during the designing step, and at the experiment level during the validation step. The strategy design process, by its nature, focuses on optimizing resources, without giving up the quality of the targeted results.

It is important that budgets and timelines are perceived as best guesses at the time they are defined and not as the absolute truth. They should be updated and communicated to decision takers, each time new insights have been gained that have a significant impact on them, either positive or negative.

### 4.1.4  Assessment of the Change Capacity of and Underlying Risks for the Firm

Assessing the *change capacity* of a firm is like solving the "chicken and egg causality dilemma—which one came first". Implementing a new or revised strategy in an organization requires it to change. But any organization can only take so much change at a given point in time. In addition, change results in disruption, which inherently increases existing and opens the firm to new business risks. Both aspects need to be well understood. Even though the strategy design process should not be primarily driven by a firm's capacity to change, understanding the boundaries towards change helps make strategic decisions that are implementable in a sustainable way.

Assessing the capacity to change of a firm, its management, its employees, and partners as well as suppliers, reverts to answering a set of questions. The answer to each of the questions should be assessed, for example, on a scale of very weak to very strong, relative to whether they inhibit or support change. Averaging the obtained scores, or even better calculating medians, quantifies a firm's perceived capacity towards change. The list of fifteen questions in Table 4.1 describes a typical set of questions to ask and answer. They focus on the five dimensions:

(1)  Relevance of change.
(2)  Emergency of change.
(3)  Speed of change implementation.
(4)  Experience with change.
(5)  Expertise with change.

These can be represented using a spider diagram as shown in Fig. 4.3.

The capacity to change assessment questions can be classified into three categories, relating to senior management, to employees, and to external stakeholders.

**Table 4.1** Sample list of questions to answer for determining a firm's capacity toward strategic change

| Questions—To Senior Management |
| --- |
| (1) What are the drivers behind changing/amending the strategy and how important are they for the success of the firm (relevance of change)? |
| (2) Is updating the strategy a top-priority issue or just something the firm feels they should deal with (urgency of change)? |
| (3) How important is it for senior management to see tangible results quickly (speed of change implementation)? |
| (4) Has the firm previously been successful in attempts to develop new or update existing strategies (experience with change)? |
| (5) How significant is the firm's knowledge around developing and implementing strategic changes (expertise with change)? |
| Questions—To Employees |
| (6) To what extent does a strategy project respond to goals employees see as important (relevance of change)? |
| (7) How enthusiastic have employees been in the past towards strategic change (urgency of change)? |
| (8) What is the employees' attention span relative to change (speed of change implementation)? |
| (9) How successful were past change initiatives from an employees' perspective (experience with change)? |
| (10) How significant is the employees' demonstrated capacity to absorb new ideas and exploit them usefully (expertise with change)? |
| Questions—To External Stakeholders |
| (11) How significant is the external pressure towards strategic change (relevance of change)? |
| (12) How eager are external stakeholders to see strategic change happen and how will they be affected by it (urgency of change)? |
| (13) How quickly do external stakeholders, especially investors, want to see tangible results from strategic change (speed of change implementation)? |
| (14) In what roles have external stakeholders been involved in strategic change in the past and what were their impact on it (experience with change)? |
| (15) What criteria do external stakeholders apply to value the success of any strategic change (expertise with change)? |

The questions should be adapted and amended to the specific strategic challenge at hand at each firm.

As with any change undertaking, a risk assessment must be performed beforehand. There exist two categories of risk to consider, that is, those inherent to the strategy design process undertaking, and those risks resulting from the outcome of the strategy design process, that is, surfacing during strategy implementation. For each risk, its severity and probability must be estimated, and possible mitigation scenarios defined. Table 4.2 illustrates some of the most common risks found in relation with developing strategies.

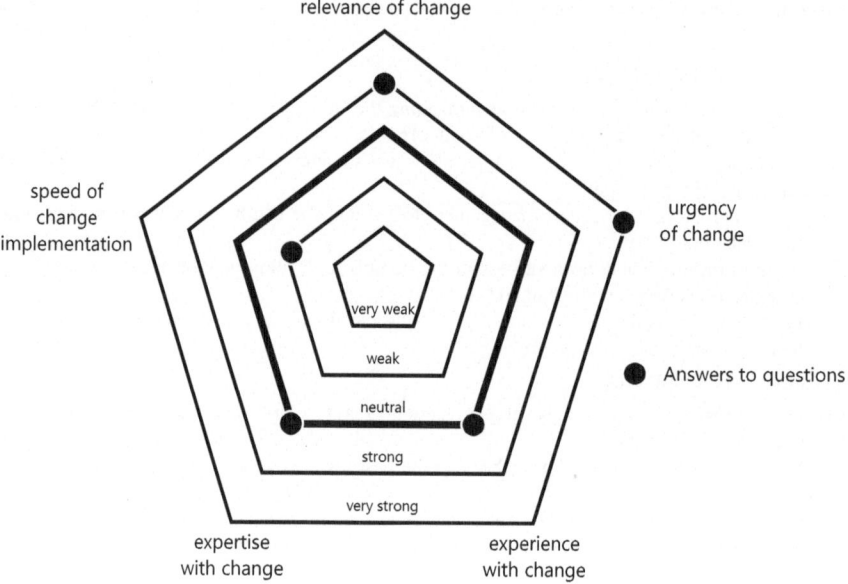

**Fig. 4.3** Spider diagram representing the firm's average or median capacity to change along five key dimensions

**Table 4.2** Sample list of some of the most common risks identified during strategy development

| Risks—Related to Strategy Development |
| --- |
| (1) Decisions taken at milestones are subsequently questioned and/or revised, unnecessarily lengthening the strategy design process |
| (2) Decision takers change during the strategy design process, leading to inconsistent decisions, making the buy-in at the end hard |
| (3) Key strategy team members leave without transferring their knowledge to other team members |
| (4) No common strategic focus can be agreed upon, failing to move to the business model layer of the strategy design process |
| (5) Too much time is spent in the observing step O, versus the designing step D, resulting in irrelevant analysis and inefficient use of resources |
| (6) Key assumptions are not validated because strategy team members believe they know better |
| (7) Assumptions to be validated are incorrectly prioritized, scheduling the testing of key assumptions that could invalidate the overall designed strategy, at the end of the validation step |
| (8) Validations take too much time and cost too much money, because an excessive level of precision is thought after |
| (9) During validation, assumptions are being marketed rather than tested, leading to biased results |
| (10) Competitors and their potential reactions are ignored while developing the overall strategy |

## 4.2  Target Industry

Before being able to initiate the strategy design process, the *target industry* must be identified. There exist two different approaches for choosing a target industry in which to compete, depending on whether taking an incumbent or start-up approach or whether starting from an existing, often already mature, business. In both cases, a sound understanding of potential target industries is required. There exist numerous approaches to acquire that knowledge, such as reading about an industry, participating in trades fares, attending conferences and seminars, or interviewing experts, to name just a few. Determining and acquiring the knowledge needed to select a given industry without wasting resources is more an art than a science and requires experience. As the strategy design process is iterative, if at a later stage, the targeted industry is found to be defined to broadly or too narrowly, its definition should be refined. Refining the target industry during the strategy design process is to be perceived as an opportunity rather than a flaw.

### 4.2.1  Incumbents

Depending on the viewpoint, incumbent firms have an easier or more difficult stance selecting a target industry—easier, as they can choose on the greenfield, harder, as there is not existing infrastructure in place. The lightweight business model helps structuring the search for an appropriate target industry. Its four components provide four different directions along with to search.

First, incumbents may select a group of customers and associated needs to be satisfied. These needs then lead to a target industry which aims at satisfying them. For example, the needs for transferring money between families in third world countries may be considered. This would lead to identifying the payment industry as target industry in the strategy brief. In general, the target industry should be defined in rather broad terms, avoiding giving up opportunities too soon. But it must be focused enough to avoid the "lost in translation" effect.

Second, focusing on a specific technology, or more broadly speaking, an invention, can serve for defining a target industry. As an example, consider the blockchain technology, providing an immutable general ledger. As the traceability of the origin and authenticity of art collectibles is a big challenge, the art authentication industry may be identified as a valid target industry applying blockchain technology. Only identifying a technology without a targeted industry will lead to phishing in the dark expeditions and is to be avoided.

Third, incumbents may select the target industry based on specific, hard to imitate capabilities they possess. Such capabilities may be the miniaturization of electric circuitry. Possible target industries could be the spying device industry or the hearing aids industry, both industries being driven by miniaturization of electronic circuitry. Another capability driven example would be incumbents specialized in supply chain management. A target industry to aim at could be grocery stores, ensuring delivery of always fresh fruits and vegetables. Amazon follows such an approach with most of its business opportunities.

Fourth and last, but not least, incumbents could focus on cost sensitive industries, or industries that would profit from increased cost consciousness. A typical example in this category would be the airline industry. EasyJet follows such a strategic direction. But, focusing on the airline industry does not necessarily mean, engaging in the discount airline business. It can also mean, becoming a supplier that allows airlines to save time or money, for example, through improved luggage handling and tracking.

A key mistake to avoid is choosing a generic type of strategy, such as a platform or a fast-follower strategy, and then identifying an industry to which to apply it. This type of reverse engineering of strategies fails more often than not.

## 4.2.2  Mature Firms

In contrast with incumbents, mature firms already compete in one or more industries. To define the target industry underlying the strategy design process, mature firms have three options to choose from, that is,

- continuing to compete within their *core industry*,
- extending their core industry by defining the target industry as a *related or adjacent industry*, and
- choosing a *new core industry* to compete in, following an incumbent-like approach.

Most mature firms select their existing industry as target industry. Unless the industry is structurally declining, for example, due to societal changes, staying with what the firm understands best is a sound choice. Staying in the same industry as in the past does not mean, that the strategy should remain unchanged. On the contrary, keeping or regaining a competitive advantage almost certainly requires changing, or at least, adjusting the existing strategy. For example, firms competing in the premium watch industry have remained in the core industry, and still re-invented their strategy time and time again.

The second option for selecting the target industry is extending the core business by moving into adjacent industries. Zook (2004) advocates this approach. A typical example of a firm having taken this approach is Microsoft, moving from operating systems, to office applications, to search engines, to developing tablet devices, up to offering cloud services. The lightweight business model helps identifying adjacent industries. Adjacent industries are those industries that share one or more components of their lightweight business model and differ, ideally be complementary, in others. A typical example are grocery stores, extending into the on-line and home delivery industry. The customer component of the lightweight business model remains largely unchanged, whereas the capabilities are extended by an on-line platform and home delivery services.

You may think that the third option, choosing a new core is not sound for mature firms. If so, think about Nokia. Nokia was founded in 1871 as a pulp mill. In the 1990 it was leader in mobile phones for retail customers. In 2014 it entered the digital health market, an industry far away from its previous core, mobile telecommunication infrastructure. More often than not, selecting a new core, different from the current one, as target industry, is chosen when the existing industry is in structural decline or significant poor management decisions have brought the firm to the verge of bankruptcy. Kodak is probably the most prominent example in this category.

## 4.3  Guiding Principles

Relying on sound guiding principles during the strategy design process is important for its success. Guiding principles summarize fundamental beliefs that need to underlie any strategy design activity. They are usually firm specific, subjective, and not verifiable. They are the strategy's axioms.[1] They are important to keep the strategy design process on the right track. They provide boundaries avoiding getting lost or getting stuck. Sometimes guiding principles are called design criteria (Liedtka et al. 2014). Ideally, they are actionable, specific, and unique No matter how great an idea will be, it will ultimately be subject to the firm's standards and principles (Mootee 2013).

Guiding principles can be classified into four different categories, that is,

- things that must be satisfied,
- things that should be satisfied,
- things that should be avoided, and
- things that must be avoided at all cost.

Guiding principles should be kept abstract and down to a minimum. Usually, two to three guiding principles per category are reasonable.

A typical guiding principle defined in the strategy brief targeting the transportation industry could be that the strategy should have a positive impact on the $CO_2$ emissions, or that the strategy must avoid any conflicts of interest with customers at all cost. Guiding principles may vary significantly from firm to firm. When developing disruptive or blue ocean strategies (Kim and Mauborgne 2005), guiding principles may not even be needed.

Even though, it is best practice to define all guiding principles up-front, they may be amended and revised over time during the tree layers of the strategy design process. If done so, it is important that they get re-confirmed by the decision takers, responsible for the final strategy result.

---

[1]An axiom is a statement that is taken to be true and serves as a premise for reasoning and arguing about strategy.

## References

Carroll, L. (1865). *Alice in wonderland.*

Govindarajan, V., & Trimble, C. (2005). *10 rules for strategic innovators.* Boston, MA: Harvard Business School Press.

Kim, W. C., & Mauborgne, R. (2005). *Blue ocean strategy: How to create uncontested market space and make the competition irrelevant.* Boston, MA: Harvard Business School Press.

Liedtka, J., Ogilvie, T., & Brozenske, R. (2014). *The design for growth field book.* New York, NY: Columbia University Press.

May, M. E. (2016). *Winning the brain game.* New York, NY: McGraw Hill.

Mootee, I. (2013). *Design thinking for strategic innovation.* Hoboken, NJ: Wiley.

Verganti, R. (2009). *Design-driven innovation.* Boston, MA: Harvard Business Press.

Zook, C. (2004). *Beyond the core: Expand your market without abandoning your roots.* Boston, MA: Harvard Business Review Press.

# A Novel Strategy Development Process Based on Design Thinking

**5**

*Design is not just what it looks like and feels like. Design is how it works*—Steve Jobs

Most traditional strategy design processes are highly analytical. They are tedious and are built on abstract concepts, like a vision, a mission, and values statements. Although such statement may be sound in communicating about strategy, they often are challenging in driving the creative process of designing a strategy. They fail to provide the necessary guidance, that any creative process requires to avoid derailment. Typical strategy design theory focuses on the capabilities and resources that define a firm's competitive positioning. Little is left to creativity, and especially creativity at the strategic level. Innovation is related to technology rather than to how to conduct business. Traditional strategy development exercises, based on deductive data driven reasoning techniques, often end up in large binders of PowerPoint presentations, and substantial consulting bills. Too much time is spent on analyzing data about markets, their size, and competitors. Too little time is used to understanding customers and their jobs-to-be-done. This does not have to be the case!

Originally, mainly architects and urban planners relied on design thinking for developing innovative solutions. During recent years, design thinking has become a mainstream wicked problem-solving approach. Based on abductive reasoning, a formal logic of inference that starts with observing and identifying the nature of the desired value to achieve and seeks simple and most likely explanations (Dorst 2015), this book presents a tree-layer iterative approach to designing sound business strategies. Through designing and validating, each layer relies on what has been observed and learned to come-up with novel and tested options. Whenever possible, the strategy design process avoids unfocused research analysis by combining exploratory and confirmatory phases in an iterative and top-down way. The goal is offering a practical, hands-on approach built on solid theoretical concepts that can be applied to disruptive start-ups as well as traditional corporations in developing or reviewing their strategies. And more important, it does not require a multi-year MBA to be understood and successfully applied.

© Springer Nature Switzerland AG 2020
C. Diderich, *Design Thinking for Strategy*, Management for Professionals,
https://doi.org/10.1007/978-3-030-25875-7_5

## 5.1   Process Overview

The strategy design process, also called *design thinking for strategy* ("DTS") process, is subdivided into three layers. Each layer is offering a specific focus in driving the development process. Figure 5.1 illustrates the three-layer approach and demonstrates how design thinking and game theory support strategy development at each stage, including intermediary milestones requiring decisions.

The *foundation layer* supports a high-level understanding of the industry and competition using an observing approach, with a focus on identifying those insights that matter most in developing the strategy. Based on the learnings, the foundation of the firm's strategy, that is, its strategic focus, is chosen. It is based on the four components of the lightweight business model and defines how the firm wants to compete and differentiate itself.

During the *business model layer*, the target detailed business model of the firm is designed and validated based on in-depth observations of customers, innovation capabilities, skills, and financial expertise, as well as the chosen strategic focus. Multiple iterations of observing, learning, designing, and validating, are usually necessary.

Once the business model has been finalized, the *competition layer* places it into the perspective of the industry in which the firm wants to compete. This is accomplished by seeking answers to Porter's five questions on strategy (Porter 1996; Magretta 2012). Game theory is used to understand if and how the designed business model works in its competitive environment. Depending on the findings, it may be necessary to refine the business model, or even the strategic focus.

Ultimately, the strategy developed is communicated in a way that allows its implementation to start. Strategy development and strategy implementation should not be mixed as they require different skillsets. The process from strategy development to implementation is not linear. Findings during the implementation phase may lead to adjustments in the development phase, and especially regarding specificities in the business model.

## 5.2   The Foundation Layer

The goal of the foundation layer is to decide along which of the four components of the lightweight business model the firm aims at competing. As stated by Porter (1985), as well as Treacy and Wiersema (1995), successful firms excel at exactly one component of the lightweight business model, while being competitive in the three others. If a firm decides to compete in more than one component, it will often fail due to the "stuck-in-the-middle" syndrome. Typical examples are failed airlines that tried to be both premium service providers and discounters. Note that focusing on a single lightweight business model dimension is only valid at the business strategy level, as it is possible to design a corporate strategy, that is, a strategy at the holding company level, where each business unit implements a different strategy

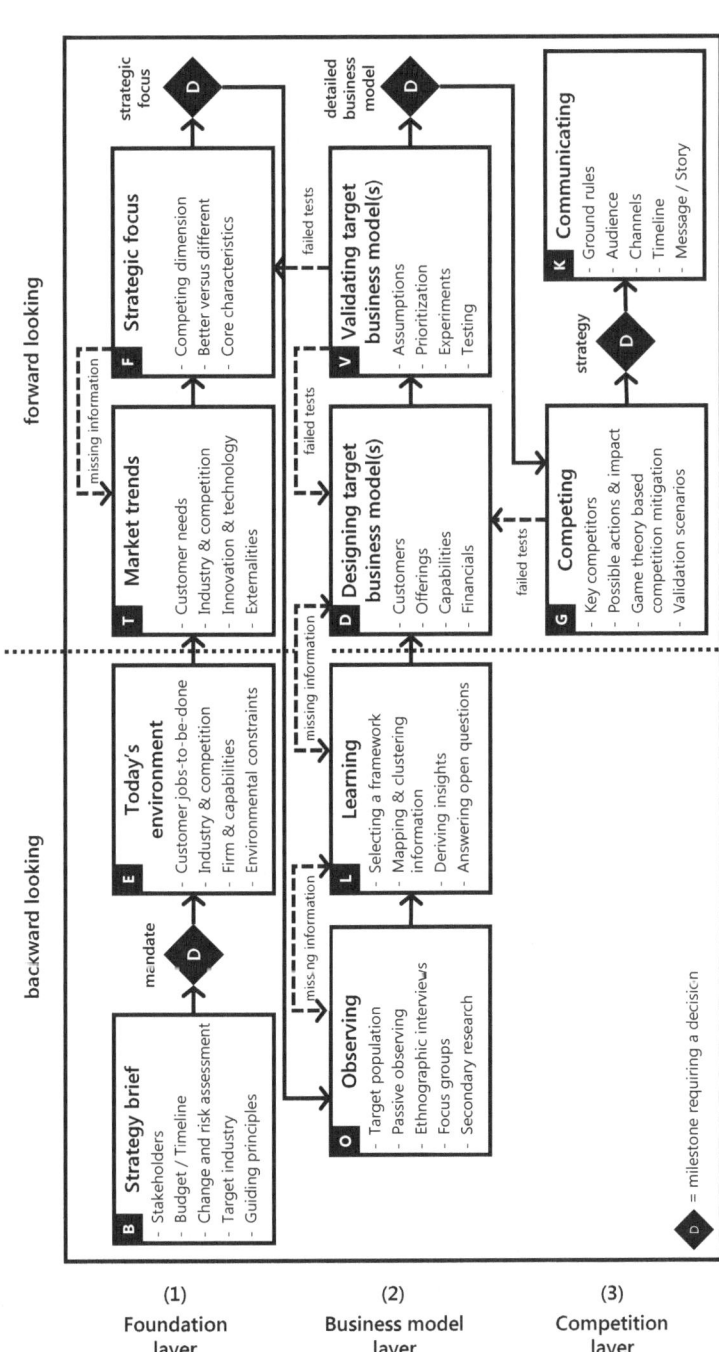

**Fig. 5.1** Overview of the strategy design process

based on a different strategic focus, that is, competing along a different component of the lightweight business model.

## 5.2.1  Strategy Brief

Before starting the design of a new strategy or update an existing one, the *strategy brief* defines the overarching scope and goal of the strategy design process, which includes

- listing all the *stakeholders* that must be involved at different points along the timeline of the strategy design process,
- implementing a *culture fostering innovation and creativity*,
- determining a *raw budget and timeline*,
- understanding the firm's *capacity to handle change* and assessing the *potential risks* underlying strategic decisions to be taken
- identifying the *target industry* in which the firm expects to compete, and
- defining the *guiding principles* on which the strategy to be developed should be based.

In contrast with common project management approaches, the strategy brief is kept short and concise to avoid unnecessarily constraining the strategy design process. The strategy brief must avoid anticipating any possible outcome.

## 5.2.2  Understanding Today's Environment

Key industry players, competitors, and the firm, are observed, and findings are documented using the lightweight business model framework. The focus is put on what matters most by applying the 80/20 rule,[1] also called Pareto principle, and not on describing every little detail. Regulatory, political, economic, social, environmental, and technological externalities are observed and documented using a separate instance of the lightweight business model.

> **Example** Consider the payment industry, focusing on the online shopping world. Figure 5.2 illustrates today's environment focusing on the four components of the lightweight business model, that is, customers, offerings, capabilities, and financials. Customers are subdivided into retailers, providing the payment services, and buyers, using the services to pay.

---

[1]The 80/20 rule, also known as the Pareto principle, states that roughly 80% of the effects come from 20% of the causes. Juran suggested the principle, and named it after the Italian economist Vilfredo Pareto, who noted the 80/20 connection in his 1896 paper *Cours d'économie politique*.

| Customers | Offerings | Capabilities |
|---|---|---|
| Retailers looking for:<br>• Solutions that integrate into their supply chain management system<br>• Solutions that are widely accepted by buyers<br>• Costs related to the attractiveness to clients served<br>Buyers looking for:<br>• Ease of use<br>• Security<br>• Acceptance by on-line stores<br>• Supported by banks<br>• Providing a credit line | • Credit card issuers, mainly large players (VISA, American Express, Mastercard, etc.)<br>• Global technology firms, offering payment solutions like PayPal, ApplePay, SamsungPay<br>• Large retailers, offering their own payment services<br>• Banks, exploiting their wire transfer capabilities<br>• Niche solutions focusing on specific markets and/or specific channels | • Interconnectivity<br>• Technologies, like NFC, Bluetooth<br>• Card transaction processing<br>• Wire routing and processing<br>• User experience design<br>• Reach, both to retailers and buyers<br>• Credit facility |

| Financials |
|---|
| • Transaction based pricing<br>• Volume based pricing<br>• Pricing power based on reach<br>• High automation, minimizing variable costs |

**Fig. 5.2** Illustrative example describing key insights of today's environment around payment services, focusing primarily on the on-line shopping experience

## 5.2.3 Identifying Industry Trends

Various design thinking tools are applied to identify key industry trends along the four dimensions

- customers,
- innovation,
- capabilities, and
- externalities.

Possible industry trends are identified by trying to extrapolate today's environment into the future. The identified trends may be inconsistent among themselves and occurrence probabilities should be associated to them. They are predictions of the future and must be considered as such.

**Example** Capability trends in the payment example shown, may include for example blockchain technology. Another trend identified may be leaning towards global offerings, focusing on a limited set of core features, rather than distinct domestic only solutions aiming at offering customized payment services, including wire transfers and mini consumer loans.

### 5.2.4  Choosing the Firm's Strategic Focus

The *strategic focus* of the firm is determined based on what has been learned from observations in the two processes understanding today's environment (process E in Fig. 5.1) and identifying market trends (process T in Fig. 5.1). The strategic focus is based on the lightweight business model component along which the firm wants to compete. It includes whether the competitive advantage should be based on being different or being superior. The other three components' key characteristics are also derived and documented as part of the strategic focus chosen.

At the end of the foundation layer, the firm should clearly recognize where it wants to develop its competitive advantage and why. The details regarding the "how" remain to be determined.

## 5.3  The Business Model Layer

The business model layer of the strategy design process aims at defining the strategic aspects which are needed by the firm to conduct business successfully. The focus is on the firm, rather than the industry, its competitors, or the external environment. Strategic business aspects are holistically addressed. The business model layer follows the four steps of the design thinking methodology, specifically, observing and learning, by looking backward, and designing and validating, by looking forward.

### 5.3.1  Observing

Rather than being unfocused, the observing process O targets observations around the strategic focus as defined in the foundation layer. Observing intends to lay the foundation for learning what customer needs are not met or met in an insufficient way and which jobs-to-be-done are relevant. Observing should not be confused with the traditional strategy analysis phase, focusing on market sizing. Passive observing aims at answering the "what" question and is often followed by interviews around the "why" questions to deepen understanding.

> **Example** Consider a hardware store that wants to re-focus its strategy along the financials dimension, notably competing to become superior in its cost management to be able to match competitors' prices. One key cost dimension is the service and support offered during the customer decision journey. Passive observing would involve identifying when and why customers seek human support, with a focus on the "what "question, that is, what do customers want to know? In a subsequent step, the observer would attempt to understand "why" customers seek human support by conducting ethnographic interviews. Is it because no alternative sources of information are available? Or is this due to a lack of understanding? Or is it even because of an emotional need for human trust?

The key to successfully observing is avoiding an interpretation of the findings, prioritizing them, or trying to find solutions to the observed pain points. Observing must be neutral and not focused only on negative aspects. Positive observations are as important.

In addition to observing based on the chosen strategic focus, the observing process should be used to gain information about externalities, such as regulatory constraints. In terms of mathematical terminology, the question to be answered is "are the constraints binding or is there still slack"? It is good practice to use focus groups for generalizing the observed insights and outcomes from interviews and conduct secondary research to identify supporting or contradicting arguments related to the findings.

## 5.3.2  Learning

While still focusing on the past, the learning process L aims not only at understanding what has been observed, but also on gaining unique insights that may be exploited towards a competitive advantage. A centerpiece of learning involves separating relevant insights from irrelevant ones. Knowledge is extracted from the observations, structured, and related to the business model's different elements. Insights move beyond the original customer- or human-centric design thinking. They also relate to non-customer facing activities, like observed capabilities, unique technologies, or distinct challenges identified when trying to address the jobs-to-be-done, including investments and expenses. Depending on the knowledge gained, further passive observations may be needed following ethnographic interviews.[2] Such iterations are a part of the strategy design process and should not be negatively connotated.

> **Example** Consider again the hardware store example. Assume you have observed that customers ask for human assistance after spending time considering various alternatives for buying a given tool and before making a final purchasing decision. Key questions asked to the human assistance relate to specific features of the tools that the customer has included in his consideration set.[3] First, it is a sound idea to use a framework to structure the information gained, in this case, using the McKinsey's consumer decision journey framework (Court et al. 2009). Knowledge is extracted from the observations by mapping the observed onto the chosen framework. This could include customers looking for comparative as well as objective information about the tool alternatives not yet identified. Customers may trust a

---

[2]Ethnographic interviews are directed one-on-one interviews, aimed at understanding the behaviors and rituals of people interacting with individual products and services. They aim at better "understanding" the jobs-to-be-done and associated pain points as well as unmet sought-after gains identified during passive observing.

[3]The consideration set is the set of products to which a person has narrowed down their choice for buying from, based on their personal screening criteria.

human salesperson to make that information available to them in an objective way. One observation may reveal whether the relevant issue is a lack of available information, its complexity, or the objectivity of the salesperson. The gained knowledge about lack of availability or mediocre quality of the information is associated with the value proposition element when mapped onto the detailed business model in addition to the specific framework chosen. Customers focus more on the human trust aspect than on the actual information would be mapped to the customer relationship elements of the detailed business model. The learned knowledge is put into perspective relative to the chosen strategic focus. Figure 5.3 illustrates possible gained knowledge mapped onto the detailed business model framework. Note that the goal of the learning process is not to identify solutions to potential issues, but to understand the jobs-to-be-done and the root causes of the identified challenges.

Ideally, the learning outcomes cover all elements of the detailed business model. They should at least cover all elements that directly or indirectly relate to the chosen strategic focus, the value proposition, and the products and services elements. For example, if the chosen strategic focus is customers, then the learned knowledge should cover the customer segments, their jobs-to-be-done, the customer relationship, and the customer delivery elements. In the case where the observations and the subsequently derived knowledge fail to provide relevant insights, additional iterations of observing and learning must be performed, or the strategic focus chosen during the foundation layer revisited.

| Customer Segments | Customer Relationship | Value Proposition |
|---|---|---|
| • Cost-conscious, but not poor<br>• Knowledgeable of characteristics sought after<br>• Rational decision maker<br>• Well informed and experienced<br>• Technology savvy | • Needs to see and touch the tools before buying<br>• Seeks advice before buying to ensure needs are effectively met<br>• Has experience with technology/app-based support | • Assumes pre-sales support is part of the offering<br>• Seeks understanding of the uniqueness of their needs<br>• Looks for cheapest in-store price within a decent travel range |
| **Customer Jobs-to-Be-Done** | **Customer Delivery** | **Offerings** |
| • Needs specific features to perform a specific job<br>• Buying decision is based on price, given needs are met | • Walk-in customer type<br>• Wants to buy and leave with the tool once they has made their decision | • Favors suppliers that offer large variety of tools adapted to specific needs<br>• Availability of tools is important<br>• Brand is not relevant |

**Fig. 5.3** Mapping observations from ethnographic interviews onto the detailed business model focusing on customers and offerings (not shown: capabilities and financials insights)

### 5.3.3   Designing

The third process of the business model layer, during the designing process D, is forward-looking. Starting with covering elements related to the strategic focus, viable options of the firm's target detailed business model are designed. The insights gained from the learning process L serve as a point of reference. Although the goal is not to restrict creativity, the designed business model options should align with the chosen strategic focus. For example, if the strategic focus is defined as competing on price, it is unsound to propose a business model option based on a sophisticated and expensive after-sales support approach for the customer delivery element. It would be preferable to implement a discounter strategy or focus on cost reducing capabilities. The trends identified during the foundation layer analysis serve as guidelines to focus the creativity and ideation during the designing process D. Similarly, externalities should be perceived as potential opportunities to be exploited, rather than restrictions.

> **Example** Consider again the previous hardware store example. One knowledge gained is that some customers seeks answers from human salespersons to specific questions related to comparing the features of the tools to buy. One design choice, given a financial strategic focus, would be focusing on the customer segments that do not require human pre-sales support. Alternatively, the firm may decide to offer customized pre-sales supports while simultaneously remaining a discount retailer. This is where real creativity is needed. An artificial intelligence-based kiosk-style pre-sales support mechanism, or even an autonomous robot, could replace humans to deliver pre-sales support to customers. This would allow avoiding excessive costs associated with relying on human personnel for pre-sales support. Another idea could be to charge for the human pre-sales support, after validating the customer willingness to pay for it.

At the end of the designing process, a complete description of the detailed business model prototype should be available. Additionally, all elements should have been checked for consistency among each other. For example, if a given value proposition is offered, it must match a given customer jobs-to-be-done element on the customer side as well as activities in at least one of the three activities elements of the detailed business model. Again, if the detailed business model's description is incomplete or inconsistent, it is necessary to reiterate the observing and learning processes, or even revisit the foundation layer.

### 5.3.4   Validating

Assumptions made during the designing process are explicated during the validating process V. They are reformulated as testable formal strategy hypothesis. Executives and strategists make too often unjustified assumptions without knowledge. Hypothesis, that sound logical on paper, often fail a "reality" test. Therefore, it is key to test all formulated assumptions underlying the designed business model options in a real-world environment. To do so, experiments must be developed and conducted. Rather than confirm assumptions using statistical theory, the goal should be on identifying what could make the assumptions fail. Hypothesis validating in strategy involves finding unexpected flaws rather than confirming the obvious. It is important to note that assumptions made for each element in the detailed business model must be validated, as well as assumptions underlying the relationships and interactions between elements. For example, if the business model designed focuses on offering a specific value proposition to a specific customer segment, it is key to validate that a customer relationship exists to links the customer segment to the value proposition.

**Example** When considering the hardware store example, the designed business model prototype assumes that pre-sales support can be provided by artificial intelligence driven kiosks or robots at a cost significantly lower than that of human product sales experts. Three key assumptions underlying this design choice are:

(1) Customers accept pre-sales support kiosk-style mechanisms or robots as an alternative to human pre-sales professionals, assuming the same level of pre-sales support quality as provided by humans.
(2) Kiosk-style mechanisms or robots, supported by artificial intelligence technology, can provide pre-sales support at a quality level that is accepted by customers as equivalent to that of humans.
(3) Pre-sales support robots can be built or bought and trained at sufficiently low cost to support the discounter's strategic focus.

Assumptions should be prioritized in increasing order of the complexity of validating and relevance to the validity of the business model prototype. A mock-up kiosk or robot could be built to test the first assumption, answering customer questions remotely by a human without the customers knowing so. This would allow testing whether customers accept kiosk-style mechanisms or robots instead of humans, at the same level of pre-sales support quality.

The validation phase aims on failing fast to succeed faster, while ensuring the detailed business model's viability.

## 5.4   The Competition Layer

Although critical, the strategic focus and the detailed business model options are only two aspects of what defines a successful strategy. A third characteristic involves determining and understanding the firm's position in its competitive environment. The competition layer of the strategy design process includes two major processes:

(1) First, the competing process G determines an understanding of how the firm aims at competing and differentiating itself from peers and prepares for potential competitor reactions.
(2) Second, the developed strategy is communicated through process K in a way that managers and employees understand it, while providing sufficient details to support the strategy's implementation.

### 5.4.1 Understanding the Competitive Landscape

At the core of understanding the firm's competitive position stand Porter's five questions about competition in strategy (Porter 1996; Magretta 2012). They are answered during the competing process G to identify the firm's competitive advantage. This means,

- identifying the *distinct value proposition* elements of the detailed business model,
- relying on a *tailored value chain* in the activity elements of the detailed business model,
- making choices or *trade-offs that differ from those of competitors* throughout the detailed business model,
- ensuring that *choices made are interdependent* and support each other, and
- offering some sort of *continuity over time.*

**Example** In the hardware store example, the distinctive value proposition as well as the trade-offs made are based on offering discounted prices combined with pre-sales advice not found at competing discounter hardware stores. The tailored value chain is supported by the use and reliance on artificial intelligence and robots to offer advice. The interdependence, or what Porter calls "fit", is ensured by including pre-sales advice in the process supporting the customer decision journey, which is primarily price-driven. As the strategy's foundation, being perceived as a discounter, does not change, the need for continuity over time is ensured.

Understanding the firm's position in the competitive landscape is important, not only for firms implementing strategies with a certain industry power, but also for those firms that aim to disrupt their market—even start-ups. Furthermore, game theory (Morgenstern and von Neumann 1947; Straffin 1993; Ghemawat 1997), including the Nash equilibrium theorem and min-max game trees, is used to understand and predict how other industry participants may react to certain strategic decisions, enabling to better understand and strengthen the competitive positioning of the firm. Understanding the challenges faced by a given positioning choice is less critical for firms implementing a blue ocean strategy (Kim and Mauborgne 2005), that is, a strategy that deliberately avoids competition through its unique positioning, than for those operating in a crowded environment.

**Example** Consider again the example of a hardware store competing on price. A key differentiator proposed in the strategy involves offering extensive pre-sales support using kiosk-style mechanisms or robots. However, reactions from competitors and their implications to this strategy must be understood before committing to it. Table 5.1 illustrates four scenarios that describe the customer's expected reactions and their implications for the firm's strategy.

Two consequences can be derived from this analysis. First, customers' perceived value of the advice is key to whether offering pre-sales advice as a discounter is sound. Second, under the assumption that a kiosk-style mechanism or robot's advice can be replicated, superior capabilities to offer advice are necessary to compete with the proposed strategy. Rather than assign probabilities to each competitor reaction scenario, the analysis is defined as a worst-case analysis.

**Table 5.1** Four scenarios reviewing how competitors may react to the discounter strategy

|     | Competitor reaction | Customer reaction | Outcome for the firm |
| --- | --- | --- | --- |
| (1) | – Only competing on price<br>– No active price wars<br>– No pre-sales support offered | – Customers do not switch at similar discounted prices<br>– Advice is perceived as a free good which some customers value | – Attracts customers seeking advice switching from competitors<br>– Customers that do not value advice do not switch |
| (2) | – Competing through price war<br>– No pre-sales support offered | – Customers that only buy on price, buy from the cheapest retailer<br>– Customers relying on pre-sales support in their buying decision process, will seek such support | – Differentiated positioning versus pure discounters possible<br>– Loses customers solely buying on price<br>– Attracts customers based on the perceived value of pre-sales support |
| (3) | – Primarily compete on price<br>– Offer similar technology-based pre-sales support | – Customers perceive advice as a free good<br>– Customers remain indifferent<br>– Some customers accept technology-based pre-sales support | – Price discrimination occurs at similar pre-sales support quality<br>– Superior technology-based pre-sales support quality attracts some customers from competitors |
| (4) | – Primarily compete on price<br>– Offer human-based pre-sales support | – Customers prefer human-based over technology-based advice | – Customer will switch to competitors with comparable price<br>– The outcome depends on the sustainability of competitors' strategy (costs) and the reluctance towards technology-based pre-sales support |

The first process step, answering Porter's five questions, is a designing step while the second step, analyzing the firm's competitive position using game theory, focuses on validation. Findings from the earlier stages of the strategy design process may lead to a reiteration of previous steps to make adjustments to the retained detailed business model and re-validate the revised strategy.

## 5.4.2   Communicating

The second process of the competition layer and the final step of the strategy design process is the communicating process K. It summarizes the outcomes of the three layers:

| Customers | Offerings | Capabilities |
|---|---|---|
| • # pre-sales support requests/sale<br>• Pre-sales support satisfaction level<br>• # switching customers | • # sales/visiting customers<br>• Market share growth | • # robots providing pre-sales support<br>• % correctly answered pre-sales questions |

| Financials |
|---|
| • Pre-sales support costs / revenues<br>• Profit margin<br>• Buying power for raw material |

**Fig. 5.4** Sample description of the business model insights used in communicating the strategy

(1) The *strategic focus* from the foundation layer.
(2) The target *detailed business model* resulting from the business model layer.
(3) The *competition insights* gained from the competition layer.

Depending on the firm's culture, a vision, mission, and values statements, may be derived and used in communications. Firms that are accustomed to key performance indicator-based strategies may summarize the developed strategy by assigning a set of key performance indicators to each of the lightweight business model elements.

> **Example** Figure 5.4 illustrates possible key performance indicators used in conjunction with the communicating process and structured around the lightweight business model's elements.

## References

Court, D., Elzinga, D., Mulder, S., & Vetvik, O. J. (2009). The consumer decision journey. *McKinsey Quarterly* (3).

Dorst, K. (2015). *Frame innovation.* Cambridge, MA: MIT Press.

Ghemawat, P. (1997). *Games businesses play: Cases and models.* Cambridge, MA: MIT Press.

Kim, W. C., & Mauborgne, R. (2005). *Blue ocean strategy: How to create uncontested market space and make the competition irrelevant.* Boston, MA: Harvard Business School Press.

Magretta, J. (2012). *Understanding Micahel Porter.* Boston, MA: Harvard Business Review Press.

Morgenstern, O., & von Neumann, J. (1947). *The theory of games and economic behavior.* Princeton, NJ: Princeton University Press.

Porter, M. E. (1985). *Competitive advantage.* New York, NY: The Free Press.

Porter, M. E. (1996). What is strategy? *Harvard Business Review, 74*(6), 61–78.

Straffin, P. D. (1993). *Game theory and strategy.* Rhode Island, NJ: American Mathematical Society.

Treacy, M., & Wiersema, F. (1995). *The discipline of market leaders: Choose your customers, narrow your focus, dominate your market.* New York, NY: Perseus Books.

# Part III
# Laying the Foundation for a Successful Strategy

# Understanding the Industry Environment and Its Implications to Strategy

**6**

*The environment is everything that isn't me*—Albert Einstein

Immediately and instinctively starting to design a new strategy or identifying changes in an existing one, almost never leads to a successful and elegant solution (May 2016). It is important to start the strategy design process by understanding the environment in which the firm aims at competing from different perspectives. There exist four key perspectives to consider:

(1) The *customers* and their jobs-to-be-done perspective.
(2) The *industry* as a whole and its participants perspective.
(3) The *firm* and its own capabilities perspective.
(4) The surrounding *environmental constraints* perspective imposed by political, economic, societal, technological, legal, and environmental circumstances.

At this stage it is not necessary to consider individual competitors separately unless they occupy a dominant position. While analyzing and trying to understand these perspectives, it is important to take both a backward looking, lessons learned, and a forward looking, trends based, stance. To do so, following combinations of exploratory observing and confirmatory learning steps, ensures that the focus is put on the most relevant insights. Observing and learning focuses on understanding as a foundation for ideation and creativity. It is not aiming at statistical or theoretical proofs. The main challenge faced is not gaining too little insights but getting lost in the wealth of information available. The goal of understanding the environment is to identify smart data rather than big data. Success depends on avoiding the over-analysis fallacy found in many of the traditional strategy development approaches.

The environmental analysis, part of the foundation layer of the strategy design process, aims at framing the foundation onto which successful strategies can be built. It provides insights to

© Springer Nature Switzerland AG 2020
C. Diderich, *Design Thinking for Strategy*, Management for Professionals,
https://doi.org/10.1007/978-3-030-25875-7_6

- steering the observing and designing exploratory, divergent thinking steps of the design thinking methodology,
- avoiding non-value-adding, useless, analysis, and
- providing supporting arguments at the different decision milestones of the strategy design process.

The industry environmental analysis should result in a concise document, or even better an A0 poster, that will accompany the design team throughout the whole strategy development process and serve as

- a continuous reality check, and
- a source for new ideas.

## 6.1   Current Environment Analysis

The *current environmental analysis* process E focuses on understanding the different dimensions impacting the strategy of the firm as of today. The goal is understanding the present, rather than predicting the future. It takes an outsider perspective to analyzing and describing the current environment. Understanding the current environment is one of the few activities in the strategy design process that can be outsourced to industry experts or consultants without having a negative impact on the overall outcome. Outsourcing may even be beneficial, as it provides a distinct and fresh perspective.

---

**Process E—Current Environment Analysis**

E.1   Understanding customers and their jobs-to-be-done
E.2   Getting a glimpse on the industry and its state from an outsider perspective
E.3   Identifying the firm's unique capabilities as perceived and valued by the outside world
E.4   Recognizing the constraints imposed by the external environment, that is, from a political, economic, societal, technological, legal, and environmental stance

---

### 6.1.1   Customers and Their Jobs-to-Be-Done

The goal of understanding customers and their jobs-to-be-done is gaining a high-level and broad understanding of who the customers in the targeted industry are, how they operate, how they define value, and how they differentiate amongst each other. The term customer is defined in a broad sense, that is, as customers to the target industry, rather than customers to the firm. Customers also include non-customers. The focus is on understanding the present, rather than on extrapolating towards the future.

Five sets of questions and their answers are at the center of the customers and their jobs-to-be-done environmental analysis step of the strategy design process:

(1)  Which are the most important customer segments in the target industry and how large are they? Which are the characteristics defining and differentiating each of the identified customer segments? Who is not yet a customer? Who are the reluctant customers? Who can be considered an evangelist customer?

(2)  What problems are customers in each customer segment trying to solve with the current offerings available? What are the customers' rational and emotional needs? What pains do they feel? What gains are they looking for?

(3)  What key characteristics, like for example, brand, advertising, trade shows, or word of mouth, define the customer relationships with the industry participants?

(4)  How do customers evaluate opportunities before making a buying decision? What key attributes, like features, design, or price, of the offerings influence their buying decision? What emotional aspects influence their decision process?

(5)  How are decisions to buy or not to buy an offering taken? Are these decisions rational, emotional, impulsive, individual, or collective, to name just the main possible characteristics? How long does the buying decision process take and who is involved in what role (end-user, gatekeeper, adviser, decision maker, saboteurs, etc.)?

Rather than identify all possible answers to those questions, the focus should be on the three to five most relevant answers to each question, focusing on covering 80% of the customers. Secondary research, like industry reports and trade show presentations, form the basis of the exploratory phase of the analysis. Interviews with experts are relied upon during the second, the confirmatory phase, of the analysis.

**Example** To illustrate a possible outcome of the customers and their jobs-to-be-done analysis, consider the mobile phone industry. Broadly speaking, the three most important customer segments that can be identified are

- *technologists*, that is, technology affine private individuals, whose needs for data services and being online all the time are significantly more important than for voice features,
- *traditionalists*, that is, mainly adult private customers who value the placing phone calls features, over using data-related apps, and
- *corporates*, that is corporate employees who rely on mobile phones to support their business activities.

**Table 6.1** Example of a backward-looking environmental customers and jobs-to-be-done analysis of the mobile phone industry

| (1) | Customer segment | Technologists | Traditionalists | Corporates |
|---|---|---|---|---|
| (2) | Jobs-to-be-done, needs, felt pains, sought-after gains | – Never miss a news<br>– Interact through group messaging<br>– Multimedia consumption (games, music, videos, etc.) | – Place phone calls<br>– Be reachable while on the move<br>– Be kept carefree | – Contact customers<br>– Access company data and tools<br>– Be reachable anytime |
| (3) | Relationship | – Brand<br>– Design<br>– Features | – Service provider<br>– Word of mouth<br>– Advice and purchase in store | – Buying through procurement<br>– Best price based, given a set of features |
| (4) | Evaluation | – Friends and feedback from peers<br>– On-line reviews | – Experience with brand<br>– Look and feel trial at point of sale<br>– Pre-sales support in store | – Request for proposal based<br>– Required and preferred features list |
| (5) | Decision | – Peer group pressure driven<br>– Design and features play a key role<br>– Looking for a perceived deal | – Traditional in-store sales based | – Price versus features trade-off |

Table 6.1 summarizes a possible outcome of the customers and jobs-to-be-done analysis.

It is important to note that the current environmental analysis does not formulate an opinion about where or how to compete in the considered industry. Its goal is defining an anchoring point onto which observations and design decisions in upcoming steps of the strategy design process can be based. It supports answering key strategic questions about how to be different from and/or better than competitors. It also helps answering the important question for which customer segments and jobs-to-be-done not to compete.

After having identified customer segments in the targeted industry, a similar analysis is performed on non-customer segments, that is, those currently not served by the industry, independent of the reason. The goal is getting a sound understanding why some individuals or corporations are not customers of the industry. The focus is on the industry and not on the individual firm. Three questions and their answers are at the center of the non-customer part of the customers and jobs-to-be-done analysis:

(1) Which are the most important non-customers of the industry? Which customers are very reluctant? Which non-customers have been customers and have turned their back on the industry's value proposition?
(2) Why are the jobs-to-be-done addressed by the industry out of scope or not relevant to the identified non-customers? What traits turned former customers into non-customers?
(3) What would have to change for non-customers to become customers of the industry?

It is important to answer the questions from today's standpoint and take an objective stance. Speculations must be avoided.

**Example** To better understand the analysis, consider again the mobile phone industry. One major non-customer segment are retirees spending most of their time at home. They have a well-defined daily routine and are often technology averse. Learning how to use a mobile phone is for them not worth the effort, as they would use it only rarely. A main reason for those non-customers to become mobile phone customers would be the abolishment of plain old telephones, that is, the replacement of traditional fix lines with modern voice-over-IP solutions. In this situation, a mobile phone may become a viable alternative to a voice-over-IP phone. Another non-customer segment are employees of companies that spend 100% of their time at the office and use voice-over-IP software, like Skype, to place calls directly from their computer.

## 6.1.2  Outsider Perspective on the Industry

After having reviewed the customer segments and their jobs-to-be-done, it is important to understand the industry as a whole, as well as the characteristics of its main participants. As with understanding the environment from a customer perspective, it is important to focus on the big picture, rather than the specificities of individual potential competitors, which may be considered at a later stage.

The industry perspective analysis starts by identifying what stage the industry participants are currently in, that is,

– *introduction stage*—low sales, high costs due to high fixed introduction costs, limited to no profits,
– *growth stage*—increasing sales, decreasing costs due to economies of scale materializing, limited profitability,
– *mature stage*—stagnating sales, focus on cost controlling, cash-cow type profitability, or
– *declining stage*—declining sales, increasing impact of fixed costs, decreasing profitability.

The industry stage analysis aims at supporting at a later stage ideating around how the industry life cycle may be exploited by the firm's strategy.

Once the overall maturity of the industry participants has been identified, the industry analysis focuses on three non customer-related dimensions, by answering the following three questions:

(1) What *value propositions* and associated *products and services* offered define the industry?
(2) Which *core capabilities*, in terms of activities, assets, labor, and skills, are at the center of delivering the identified value propositions?
(3) Through what *revenue generating mechanisms* is the viability of the offerings ensured?

These three questions are tightly related to the desirability, feasibility, and viability questions that are at the core of any strategy (Brown 2009).

In addition to these business model related environmental questions, a fourth question needs to be addressed:

(4) How are industry participants *differentiating themselves from each other* and what makes the different value propositions offered unique?

Although it would be possible to conduct a full-fledged five-forces analysis using Porter's framework (Porter 1979) to answer question four, the added value at this stage of the strategy design process would be limited.

Rather than answer the four industry environment analysis questions in an abstract setting, the strategy design process focuses on those specificities that matter most. Answers should relate to the industry participants with whom the firm potentially competes, that is,

- the three largest industry participants, in terms of market share,
- the three industry participants that exhibit the highest growth rate over the most recent business cycle, usually three to five years, and
- the three industry participants that show the most disruptive innovation capabilities.

If the industry is very fragmented, it may be necessary to consider more than three industry participants in each segment.

Once key players have been identified, the four questions are answered, by focusing for each question on the most prominent answers. Out of all the answers, considering commonalities, the three to six most prominent characteristics are filtered out. This leads to three documented lightweight business models, excluding customers, one per segment of industry participants, combined with competition characteristics.

### 6.1.3 The Firm and Its Capabilities

The goal of the firm's capabilities analysis is getting an objective high-level description of how the firm operates, what are its key traits, along the four dimensions of the lightweight business model. The analysis is not to be confused with the traditionally used SWOT analysis[1] (Barney 1991). Subjective interpretation is misguided. The capabilities analysis is fact-based backward looking, rather than speculative forward looking.

If the firm in focus is a start-up that has yet to design its first complete strategy, most of the firm's capability analysis can be ignored. In addition, firms that aim at re-inventing themselves, moving their business into a new core area (Zook 2004) or following a blue ocean strategy approach (Kim and Mauborgne 2005), gain only limited insights from a firm's capabilities analysis and thus can reduce it to minimum or ignore it altogether. For each of the following four questions, aligned with the four dimensions of the lightweight business model, the three to six predominant answers should be identified:

(1) *Customers.* Who are the most important customer segments in terms of revenue and profitability? A $2 \times 2$ matrix, representing the revenues of the $x$-axis and profit on the $y$-axis, supports structuring the data. The focus is on customer segments, that is, customer groups that have similar needs which are satisfied by similar offerings, rather than individual customers.

(2) *Offerings.* What are the most important offerings produced for the customer segments identified, and what is the value created for customers? A $2 \times 2$ matrix, using on the $x$-axis the number of units produced and, on the $y$-axis, its strategic relevance to the firm, allows structuring the information.

(3) *Capabilities.* What unique capabilities, in terms of activities and processes, as well as skills and knowledge, does the firm have? The focus should be on the capabilities that are unique and critical to producing the most important offerings identified in the previous question? It is also relevant to identify capabilities that the firm possesses which are not fully exploited or not used at all. This may be un- or under-used assets, patents, unique knowledge, capital, or human resources.

(4) *Financials.* How is the firm generating profits? The focus is on the cash-flow generating mechanisms, rather than on specific numbers.

In addition to understanding the capabilities of the firm, it is also important to identify negative aspects. This means, identifying what the firm explicitly does not characterize and understand the reasons behind the negative insights. The focus is on explicit decisions rather than random outcomes. The associated questions to answer, focusing on the firm and not the industry, are:

---

[1]SWOT analysis is an acronym for a strengths, weaknesses, opportunities, and threats analysis and is a structured planning method that evaluates those four elements of an organization.

(1) *Non-customers.* What customer segments is the firm explicitly not serving and why? For example, RIM does not target teenage customers using their mobile phone to play games and use social media tools. RIM defines itself as a business-to-business firm.

(2) *Left-out offerings.* Which offerings that have been identified in the industry environment analysis, is the firm not offering and why? Although less obvious, most private banks do not offer their clients tax advice. The reasons are that third-party certified tax experts are better at providing independent tax advice and that private banks do not want to onboard the legal risk associated with tax advice.

(3) *Missing capabilities.* What capabilities, skills, and resources, does the firm explicitly not possess? For, example, a generic drugs pharmaceutical company explicitly renounces on expensive new drug development research and development capabilities.

As for other analysis, the focus should be on the three to five most relevant answers. The classical Pareto principle[2] or 80-20 rule applies.

To successfully perform a firm's capabilities analysis, two traits need to be considered. First, senior managers and decision makers covering all areas of the firm must be involved. This means, individuals with knowledge in marketing, sales, customer relationship management, product development and management, production, operations, support services, finance, human resources, as well as legal and compliance, need to provide input data. Second, to avoid a biased outcome, the analysis process should be led by an external facilitator. Rather than perform individual analyses, covering the different elements of the firm's value chain, best results are achieved by running three workshop sessions structured as follows:

(1) A first exploratory session is run with a diverse team of representatives from all areas of the value chain to identify possible answers to the four positive and three negative questions. A holistic view of the firm, considering the different areas of expertise, is taken.

(2) A second confirmatory session aims at identifying and selecting the most critical answers. To allow reflection, it is best practice to separate the exploratory and the confirmatory workshops by one to two weeks. Rather than rely on abstract analytics to prioritize the gained insights, using support of a moderator offers best results.

(3) Although not strictly necessary, the outcomes of the confirmatory session should be reviewed and validated in a third session with the firm's board of directors, executive or management teams. In addition to getting validation feedback, this third session supports buy-in through fostering a shared understanding.

---

[2]The Pareto principle (also known as the 80/20 rule) states that, roughly 80% of the effects come from 20% of the causes.

## 6.1.4 Environmental Constraints

Most strategy practitioners consider the fourth perspective, the surrounding environmental constraints analysis as the least rewarding one. It extracts constraints that limit the strategy design freedom. And, who likes to be constrained. But, understanding constraints and their raisons d'être can lead to new opportunities and differentiators.

There exist different frameworks for scanning the external environment for constraints. One of the most popular approaches is the PESTLE framework[3] (Aguilar 1967). Although designed to be a part of traditional analytical strategy development processes, understanding its six factors is relevant in any strategy design process.

### 6.1.4.1 Political
The political factor supports understanding the relationship between the government, the legislator, and the industry. Key insights are free trade agreements, market access, taxes, tariffs, and labor laws, to state the most important ones. Understanding the stability of the government, as well as its democratic structures, especially the judiciary one, are important.

> **Example** The taxi disrupting firm Uber faced challenges with its business model in numerous jurisdictions because it failed to understand the difference between independent contractors and employees from a tax and social security contributions perspective (assuming this was not an intentional business model design decision). This led Uber to being fined and having to pay arrears that limited the profitability of the strategy being a technology ecosystem platform for drivers, rather than a taxi or transportation company.

### 6.1.4.2 Economic
Rather than focus on an individual or a specific industry, the economic factor identifies key parameters common to all market participants. The most important parameters are interest rates, inflation expectations, real economic growth, wage growth, consumer confidence, unemployment rates, and foreign exchange rates. They determine, amongst others, the cost of capital, and as such, the viability of capital intense strategies. The environmental constraints analysis focuses on the dependencies between the industry as a whole and the firm in particular and key economic variables. If cross-border activities are relevant, for example, because of imported raw materials, the relative strengths of involved industries, in terms of purchasing power, need to be considered.

> **Example** When developing a business model for a low-cost airline, one of the key economic challenges to be identified and addressed is the variability of kerosene prices. Poorly understanding and managing the economic driving forces behind oil prices may lead to a non-sustainable strategy.

---

[3]The PESTLE analysis (political, economic, socio-cultural, technological, legal, and ecological) is a framework based on macro-environmental factors used for scanning the environment.

### 6.1.4.3  Societal

The societal or socio-cultural factor identifies relationships between the society as a whole as well as its members and the value propositions offered. For example, the industry value proposition may require a certain degree of education or a specific mindset, like safety awareness, to be considered. Religious and cultural aspects that play a role should also be identified during the societal environmental analysis.

> **Example** A typical observation related to the societal environment is the aging population of western societies. But more important, in terms of strategy insights, is the degree of unwillingness of young people to finance the retirements of the older generation through current pay-as-you-go systems. Another societal challenge observable is the increasing fear towards globalization and the perceived loss of control.

Although many societal trends have a negative co-notation, it is the goal of the strategy designers to re-interpret them in a neutral context rather than look at them as threats.

### 6.1.4.4  Technological

Technology plays a key role in certain industries to deliver upon the promised value proposition. For example, lithium battery technology is a key technology in the electrical car industry. During the technological environmental analysis, the focus should be set of those technologies that support the industry, rather than define it. Recent innovations and their impact on quality, risk, and costs, should be identified. The technological analysis should include missing and immature technologies. When identifying the technology environment, it is key to primarily focus on existing technologies that may or may not yet have a business application. At this stage, the focus should be on the technology itself rather than its application.

> **Example** Typical technologies stemming from the IT side are the blockchain general ledger, smart contracts, cloud computing, internet-of-things, and robots. The technological analysis should go beyond pure software and hardware technologies. It should also include, depending on the target industry identified in the strategy brief, technologies like gene technology or energy transformation, transportation, and storage technologies.
>
> The challenge with the technological environment analysis, as is the case with the other analysis, is getting the breath and depth of the analysis right. Strategy designers must learn from experience how to most effectively deploy their analysis resources. Often, less is more.

### 6.1.4.5  Legal

The legal factor identifies those laws and regulations that impact what value proposition, to whom, and how it can be offered, how it can be produced, and how it can be priced. During the legal environmental analysis, all four dimensions of the lightweight business model need to be covered. Experience has shown that the legal analysis is the largest and most critical part of the environmental analysis. Key dimensions are safety, health, and industry-specific regulations, as well as customer protection laws. Most, if not all industries, are affected in one way or the other. In contrast with common belief, Murray (Cross 2011) regards regulations as a fundamental necessity to innovation.

**Example** Strategists focusing on the financial services industry must understand the MiFID regulations and their implications on the customer segments served, the offered value proposition, the handling of the customer relationship, as well as the chosen customer delivery approach. Although, for example, EU regulations allow funds registered in one country to be sold in all other EU countries using the so-called passporting rule, some jurisdictions provide advantageous options, for example with respect to tax treatment, that others do not.

**Example** Another example from the legal environment category is the adherence to standards, like sanitary standards when professionally renting rooms, as does Airbnb.

**Example** A typical legal challenge that many internet-based firms face is correctly accounting for VAT and other sales taxes. A sound understanding may allow finding a cost- and resource-efficient set-up.

### 6.1.4.6   Ecological

The last but not least important environmental topic is the ecological topic. Key aspects of the ecological analysis are identifying factors, like $CO_2$ emissions, available sunlight, or days of rain, that have a material impact on the success of the value proposition offered by the industry. Ecology goes way beyond the Paris climate accord.

**Example** An outdoor tourist attraction company needs to understand the weather in the region it operates to tailor its value proposition.

**Example** A tire manufacturer must understand how temperatures evolve over time and identify the implications on the tire treads.

---

Collins (2010) has introduced a graphical framework, called the PESTLE-Web$^{TM}$, illustrating not only the findings from the environmental analysis, but also exhibiting the relationships amongst them. This framework can easily be combined with the lightweight business model as used during the jobs-to-be-done and industry analysis.

**Example** Figure 6.1 illustrates a subset of a possible environmental analysis of the retail lending industry, a segment of the financial services industry.

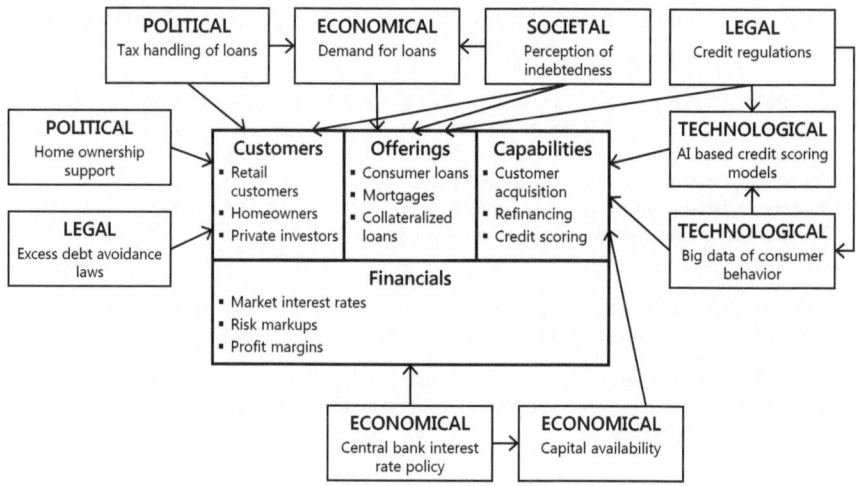

**Fig. 6.1** Environmental analysis of the retail lending industry based on the PESTLEWeb™ framework

## 6.2  Industry Tends

Up to now, the focus of the environmental analysis has been on understanding the current industry environment from four different perspectives. The analysis would not be complete without looking into the future and identifying key industry trends. Whereas the current environment analysis is objective, the industry trends review introduces a subjective bias. This is not per se good or bad. To avoid unproductive hypothesis testing of identified industry trends, the trends should only be considered, if they may be related to the firm's lightweight business model and thus being relevant during the designing process of the strategy design process, both at the foundation as well as at the business model layer. Identified trends may be complementary, or event inconsistent with each other, as they are subjective forecasts of the future.

The industry trends analysis of the strategy design process follows a similar structure as the environmental analysis does. It focuses on four complementary steps, targeting customers, the industry, technologies, and the external environment.

**Process T—Industry Trends Analysis**

T.1   Identifying changes in customer segments and recognizing unmet needs, both rational and emotional ones, new pains, and new sought-after gains

T.2   Determining changes in the industry structure, e.g., customers, competitors, suppliers, substitutes, regulators, influences, or saboteurs

T.3   Understanding trends in innovation and technology that may be relevant to the targeted industry and customers

T.4   Grasping potential impacts on industries, the industry, competition, and customer behavior

## 6.2.1   Customers

Key questions to be addressed during the customer tends analysis are:

(1)   How are customer needs expected to evolve and change in the future?

(2)   What will be the impact of demographics, education, environmental, societal, or political changes on customers and their jobs-to-be-done?

(3)   What are the driving forces that will make the industry change based on customer expectations?

The goal of the forecasts is extrapolating the customer journey of today into the future. The environmental analysis may serve as a starting point.

## 6.2.2   Industry Structure

During the industry structure analysis, changes in the industry structure are identified:

(1)   Will the industry be growing or shrinking in the future? Towards what maturity stage, that is, introductory stage, growth stage, consolidating stage, or declining stage will the industry evolve? What are the reasons behind the expected changes? How will the *J*-curve evolve (Bradley et al. 2011)?

(2)   How will the core value propositions that are currently offered change over time? Is the industry becoming broader or narrower?

(3)   How is price sensitivity expected to change? Is the industry moving towards commoditization?

### 6.2.3   Innovation and Technology

Although innovation is often used as a buzzword, the innovation trends analysis aims at identifying how the maturity of the industry is expected to evolve:

(1)  What are key innovations in recent years that may have a significant impact on the industry in the future? What trends are academic researchers currently following-up upon?
(2)  How innovative, in terms of quality and quantity of innovations, has the industry been? Were innovations rather incremental or more of a disruptive nature?
(3)  What impacts is the industry expecting from generic technology trends, like artificial intelligence, robotics, or internet-of-things?

### 6.2.4   Externalities

Key trends with respect to externalities can be identified by answering the following questions:

(1)  How is the industry impacted by globalization?
(2)  Which regulatory changes are expected in the future?
(3)  What changes in the society, like aging population, that are expected, will have a material impact on the industry?
(4)  How will the value chain of the industry change over time and which deconstructing trends can be identified?
(5)  What political trends and policies are expected to change the industry?

### References

Aguilar, F. J. (1967). *Scanning the business environment*. New York, NY: The Macmillan Company.
Barney, J. B. (1991). Firm resources and sustained competitive advantage. *Journal of Management, 17*(1), 99–120.
Bradley, C., Hirt, M., & Smit, S. (2011). Have you tested your strategy lately? *McKinsey Quarterly.*
Brown, T. (2009). *Change by design: How design thinking transforms organizations and inspires innovation*. New York, NY: HarperCollins Publishers.
Collins, R. J. (2010). *Is there a better way to analyse the business environment*. Masters Thesis, Reading, United Kingdom: University of Reading. http://users.ox.ac.uk/~kell0956/docs/pestleweb_thesis.pdf.
Cross, N. (2011). *Design thinking*. London, United Kingdom: Bloomsbury Academy.
Kim, W. C., & Mauborgne, R. (2005). *Blue ocean strategy: How to create uncontested market space and make the competition irrelevant*. Boston, MA: Harvard Business School Press.
May, M. E. (2016). *Winning the brain game*. New York, NY: McGraw Hill.
Porter, M. E. (1979). How competitive forces shape strategy. *Harvard Business Review, 57*(2), 137–145.
Zook, C. (2004). *Beyond the core: Expand your market without abandoning your roots*. Boston, MA: Harvard Business Review Press.

# Choosing a Tangible Strategic Focus Rather Than Building Upon an Abstract Vision

7

*The essence of strategy is choosing what not to do*
—Michael Porter

A key challenge faced by many strategy development and review initiatives, whether conducted internally or supported by external consultants or experts, is where to start. Traditional strategy research (Collins and Porras 1996) suggests starting by formulating a vision and a mission, describing the core ideology and envisioned future. Others, such as Grant (1991), suggest taking a resource-based approach, focusing on capabilities as the foundation for competitive advantage. More recently, authors such as Zott and Amit (2013), Zott et al. (2011) or Christensen et al. (2016), argue that the design of any strategy should start with questions related to customer needs or jobs-to-be-done. Many practitioners, including major strategy consulting companies, applying deductive approaches, extensively rely on strategy analysis tools (Harris and Lenox 2013), such as Porter's five forces, the SWOT analysis, value chain analyses, firm capability analyses, or strategy maps, to name just a few. A lot of effort is put into unfocused analysis.

The three-layer strategy design process presented in this book starts by defining the field of play through the concept of *strategic focus* of the firm. The strategic focus defines the primary dimension along which the firm wants to compete and differentiate, aligned with its targeted customers, its capabilities, and the industry environment. The possible dimensions along which the firm expects its competitive advantage to play out are defined by the four dimensions of the lightweight business model, that is, customers, offerings, capabilities, and financials. Indeed, according to Porter (1985), a firm can gain competitive advantage either through cost leadership, focusing on the financials dimension, or through differentiation. By introducing the strategic value disciplines model, Treacy and Wiersema (1995), extended the notion of differentiation by arguing that any successful firm is required to excel along one of the three dimensions customer intimacy (related to the customer dimension of the lightweight business model), product innovation (related to

© Springer Nature Switzerland AG 2020
C. Diderich, *Design Thinking for Strategy*, Management for Professionals,
https://doi.org/10.1007/978-3-030-25875-7_7

the offerings dimension), or operational excellence (related to the capabilities dimension) and be good at the other two dimensions.

## 7.1  Deriving the Strategic Focus Using Design Thinking

Process F determines the strategic focus that the firm targets as its primary dimension along which to compete and differentiate. It is abductive in nature and applies the design thinking methodology.

**Process F—Strategic Focus**

F.1  *Observing and learning.* Identifying the strategic focuses currently prevailing in the targeted industry and describing their characteristics using the outcome of the environmental analysis

F.2  *Designing.* Designing possible strategic focuses for the firm and defining the characteristics supporting them

F.3  *Validating.* Validating the designed strategic focuses by formulating and testing hypothesis

F.4  Selecting one of the designed and validated strategic focuses as the target strategic focus of the firm

Process F relies on the environmental analysis, especially the perspective on the industry, during the observing and learning steps. As shown in Fig. 7.1, based on the learned insights related to how firms compete in the targeted industry, possible strategic focuses for the firm are identified, and their characteristics summarized.

Prototyping techniques are used to design one or more potential strategic focuses for the firm. These prototypes are characterized by how the strategic focus contributes to defining a competitive advantage.

Once strategic focus prototypes—at this stage they are only prototypes—have been defined, underlying hypothesis are formulated. Hypothesis make the assumptions on which the prototypes are based explicit. They are validated or refuted through well designed, quick and cheap to perform, and easy to understand, experiments. Depending on the outcome, the designed prototypes for the strategic focus are either retained, discarded, or amended. In the latter case, the design thinking process iterates through the designing (F.2) and validating (F.3) steps. During validation (F.3), the strategic focus prototypes are challenged, and subsequently improved upon, until there are no more pending uncertainties that would fundamentally question their design (Cross 2011). The goal of validation is identifying potential flaws early, rather than confirm what is already known to be true.

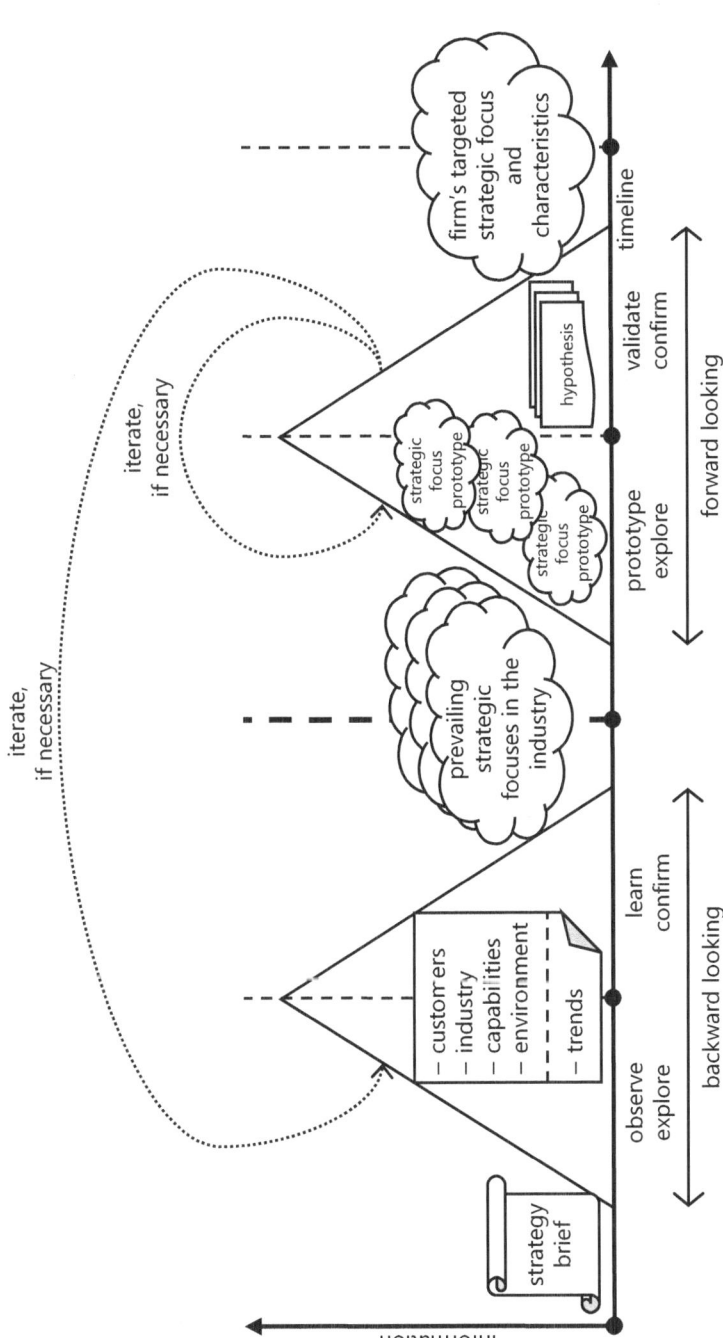

**Fig. 7.1** Design thinking process to develop the validated strategic focus and its characteristics targeted by the firm

If, during either the designing or the validating steps, insights from the observing and learning steps are unclear, or missing, the process iterates back to the observing and learning step (F.1) to gather the missing information and insights. This iterative approach, which is at the core of design thinking, ensures that the resources, that is, time and money, are used wisely.

Finally, in the selection step (F.4), the most promising strategic focus is selected out of the retained prototypes, as the firm's target strategic focus. It defines the field of play or foundation on which the new or revised strategy will be built during the business model and competition layers of the strategy design process. Having defined a solid foundation allows significantly reducing the strategy development time and increases the quality and thus the chances of success of its outcome.

## 7.2  Observing and Learning

The prevailing strategic focuses currently relied upon in the targeted industry defined in the strategy brief, are identified before starting the prototyping of possible firm-specific target strategic focuses. Insights gained during the environmental analysis are studied and the most relevant strategic focuses of competitors described. For each strategic focus detected, the following information is identified:

(1) Does the strategic focus aim at outperforming the industry by being *superior* or by being *different*?
(2) What are the three most important *characteristics* subsuming the traits of the strategic focus?

It is a good idea to label each identified strategic focus by describing the competitors in a persona-like way (see Chap. 8 for insights on personas). Knowing how industry participants compete and try to position themselves is important to ensure that the to be designed strategic focus can be a successful foundation for the strategy.

**Example** Table 7.1 illustrates three dominant strategic focuses identified in the Swiss private banking industry.

**Table 7.1** Three most common strategic focuses found in the Swiss private banking industry

| Persona | – Traditional global private bank | – Fund distribution bank | – Entrepreneur's private bank |
|---|---|---|---|
| Strategic focus | – Customers | – Offerings | – Customers |
| Competition type | – Being superior | –Being superior | – Being different |
| Key characteristics | – Service quality<br>– Personalized offering<br>– Global presence | – Large product shelf<br>–Best in class approach<br>– Driven by the CIO's market views | – Focus on specific customer segments<br>– Entrepreneurship approach<br>– No own production |

## 7.3   Designing Possible Strategic Focus Prototypes

Designing and prototyping possible strategic focuses is where shaping the firm's future strategy really starts. Although the process to be followed is systematic, its content depends on the creativity of the strategy team members shaping it.

A successful strategy is characterized by exactly one well defined strategic focus. Choosing two or more strategic focuses as the foundation for one strategy[1] leads to the "stuck in the middle trap[2]". During the design step, multiple strategic focus prototypes should be designed, and their validity explored. It is recommended to always develop more than one strategic focus prototype. Nevertheless, quality is more important than quantity. At the end of the foundation layer, exactly one strategic focus must be retained. Before moving on to the business model layer of the strategy design process, the strategic focus selected must be confirmed by the decision takers, that is, the board of directors, the executive committee, the CEO, or any other body or individual responsible for the firm's strategy. In the unlikely event that during the business model layer, or even the competition layer, a fundamental flaw is identified in the chosen strategic focus, its characterization may be refined, iterating through the foundation layer steps of the strategy design process, or even a completely different strategic focus selected.

### 7.3.1   Identifying Possible Strategic Focuses

Proposing the strategic focus to be targeted by the firm is subjective. There is no a priori right or wrong choice. The proposal is guided by the strategy brief (Chap. 4), the environmental analysis (Chap. 6), and the prevailing strategic focuses in the targeted industry. If the goal is to design a strategy relying on radical change, a disruptive strategy, choosing a strategic focus different from those prevailing in the industry is recommended. If, on the other hand, the goal is to introduce an incremental update to an existing strategy, relying on the current strategic focus or a strategic focus close to the current one is sound.

#### 7.3.1.1   Customers
Selecting customers as the strategic focus, the lightweight business model dimension along which to excel and create a competitive advantage, means putting the customers and their explicit and implicit needs first. Success depends on understanding the customers' jobs-to-be-done, their needs, felt pains, and sought-after

---

[1]The term strategy as such applies to a single firm, a business unit, or a brand within a business unit.
[2]A firm is said to be "stuck in the middle" if it does not offer a distinct value proposition that attracts customers. Stuck in the middle firms usually offer multiple mediocre value propositions that customers are unable to identify with or distinguish between and as such are not attracted by them.

gains, better than competitors. Customers are put at the forefront of any strategy decision. Reverting to Ford's quote on customers wanting faster horses rather than cars, if asked, companies with a customer strategic focus would aim at delivering faster horses, or offerings based on horses that speed-up travel. Just caring about customers, providing a superior customer experience, or listening to customers, is not enough in a customer based strategic focus. Value for customers needs to be created.

> **Example** Typical companies relying on a customer strategic focus are premium airlines, such as Singapore Airline, coffee shops, such as Starbucks, or family offices, such as the Fremont Group. Hilti, the tool manufacturer from Lichtenstein, most recently followed a customers based strategic focus transforming its strategy from selling tools to selling service contracts that ensure that the craftsmen have the tools they need at hand when they need them.

Many firms fail because they believe they implement a customer based strategic focus, although they focus on an offering or capability based one, putting the customer second, behind the offerings or the capabilities underlying the offerings. Being customer centric, is much harder that it may be perceived at first. Disruptive strategies are rarely customer centric as disrupting means focusing on offerings that customers are not yet aware of.

### 7.3.1.2  Offerings

At the core of any offerings-based strategic focus are novel products or services including novel features. Predicting what customers may value in the future is critical to success. Inventions and innovations are at the center of the stage. This does not mean that customers can be ignored. It means that the strategy is driven by offerings, putting customers in a supporting role, rather than a leading one. First movers typically chose to follow an offering based strategic focus. Following-up on Ford's quote on customers wanting faster horses rather than cars, if asked, firms adhering to an offering based strategic focus would invent a car, a motorcycle, a helicopter, or any other individual transportation means. Offerings based firms create needs for their products and services that customers have not thought of in the first place. Choosing offerings as the strategic focus is often a high-risk strategy, providing a high reward, if successful.

> **Example** A typical example of an offering focused firm was Apple under the leadership of Steve Jobs.

### 7.3.1.3  Capabilities

The capabilities based strategic focus is the most common one chosen by firms to design their strategy upon. Capability based firms exhibit superior skills and/or assets and/or are able to exploit them is a superior or distinct way. A capabilities-based strategic focus is typical for companies relying on the resource-based theory of strategy development (see also Chap. 1). They leverage their resources to provide a competitive advantage and deliver superior value to its customers. Capabilities based firms are often fast followers, copying new offerings from competitors, adapting them to their customers' needs, and delivering them through leveraging their superior capabilities. In Ford's quote on customers wanting faster horses rather than cars, if asked, firms implementing a capability based strategic focus would leverage their skills in breeding horses, or, come to up with different animals that can transport people and are faster than horses. Capability based firms often focus on incremental improvements, rather than radical change. Capability based strategies are common is industries that provide non-assembled goods (Utterback 1994).

> **Example** Large asset management firms, such as Blackrock, but also niche players, such as Fisch Asset Management, follow a capabilities-based strategic focus offering a portfolio of distinct products based on the same investment capabilities of the firm. An example of a disruptive capability-based strategy was the entrance of Nucor into the United States steel market, competing on implementing mini-mill processes.

### 7.3.1.4  Financials

Although the financials strategic focus is often related to discounter strategies, that is, strategies competing on price, this is not the only reason for choosing a financial strategic focus. Firms targeting a financials strategic focus typically excel at managing costs. More often than not, they differentiate themselves through different and novel pricing models. For example, firms following the financial strategic focus may excel at transforming one-off payments into recurring streams. Alternatively, revenues may be related to value delivered rather than costs incurred. For example wealth management product prices could be related to investment performance, rather than the efforts incurred by managing portfolios. If Ford would have applied his quote related to customers wanting faster horses rather than cars, if asked, to the financial strategic focus, he may have sold three horses for the price of two or may have leased the horses rather than sold them, or even charged for the time the horses took to get from A to B as a measure of performance. Financial strategic focus-based firms are typically competing in commodity industries. The more interchangeable the offerings are, the more important the price becomes, and a financials-based strategic focus prevails. Differentiating through capital availability or access to capital at a cost that competitors cannot match, is another financials based strategic focus option for a firm to compete on.

### 7.3.2  Choosing How to Compete

Once a strategic focus has been determined, the question whether competing though superiority or differentiation needs answering.

(1) *Superiority*. The firm competes by being better than its competitors, for example, by delivering superior products or services, providing better pre- and after-sales support, or excelling at execution through quality or speed.

(2) *Differentiation*. The firm competes by being different when compared to competitors. Differentiation may be through any element of the business model that is related to customers. Differentiation only exists if customers perceive it as such.

Unless the firm competes on commodity offerings or is in a buyer driven industry,[3] attempting to be superior without being different rarely works.

### 7.3.3  Characteristics Supporting the Strategic Focus

Depending on how to compete, there exist different approaches for identifying the key characteristics supporting the chosen strategic focus. The identified characteristics should be limited to the most important ones. Ideally, the characteristics are described by bullet point lists. Alternatively, graphical illustrations may be used. Supporting prose makes it easier to formulate validation hypothesis later on. There is no need for more in-depth insights at this stage.

If aiming at competing through being superior, the characteristics underlying the chosen strategic focus should be derived from those of competitors or of the targeted industry. Whether or not the firm has a chance to compete successfully through being superior will be determined during the strategic focus validation step.

**Example** Typical examples of firms competing on superiority, superiority being defined as cheaper, are discounter grocery stores such as Aldi or Lidl.

When a firm chooses to compete by being different in one or more areas around its strategic focus, the unique characteristics supporting differentiation need to be identified. For example, Starbucks, following a customer based strategic focus, differentiates itself by offering an atmosphere where customers can linger without being pushed to consumption. Other characteristics, such as engaging customers through a loyalty program, or offering highly customized beverages, are shared with competitors.

Note that it is possible that different firms show similar characteristics at the strategic focus level, without being identical or competing through being superior. Differentiation may also come from a distinct combination of superiority traits. The details of the differentiating elements will be designed into the strategy at a later stage, during the business model and competition layers of the strategy design process.

---

[3]A buyer driven industry is an industry in which the buyer, rather than the firm, dictates the strategic focus the firm must follow.

**Example** To illustrate how prototyping allows designing a possible strategic focus, consider a small independent fund management boutique currently implementing a niche strategy around producing actively managed value based mutual funds and distributing them through third-party solution providers, such as private banks.

Figure 7.2 illustrates (top left) the firm's current offerings based strategic focus. Based on the industry analysis, a summary strategic focus, aggregating key competitor insights, is derived and shown in the top right in Fig. 7.2. Both the firm and the targeted industry implement an offerings strategic focus. This is not uncommon with niche players or boutique firms. Indeed, many boutiques exist solely because their founder had an innovative idea. It is only when scaling the business that alternative strategic focuses become an option.

Looking at the customer segments served by the firm and the customer segments identified at the industry level, it becomes obvious that achieving a competitive advantage by focusing on a customers strategic focus, would be sub-optimal, as the firm would have to invest in building direct access to a customer base (rather than distributing through 3rd party solution providers). Another alternative would be to shift the focus onto the institutional investors customer segment. Both options can be discarded without extensive research and analysis, based on the simple observation that retaining a customers strategic focus would be in contradiction with being a small independent boutique.

Although the firm has unique capabilities with its in-house value research and efficient outsourcing operations, building a competitive advantage on a capabilities strategic focus, is discarded because the market values other capabilities higher, like indexing or storytelling. Economies of scope, that is, offering a portfolio of similar funds, all reverting onto in-house value research, are hard to realize under a boutique structure. Abandoning the boutique structure can be discarded due to ownership and associated capital constraints, two guiding principles identified in the strategy brief. All those decisions can be taken based on the limited amount of information derived during the environmental analysis step of the strategy design process.

Comparing the financials component of the firm with the characteristics of the industry does not show much room for differentiation either, as becoming a cost leader is not a viable option due to a lack of economies of scale under a boutique structure.

Finally, the firm should opt for an offerings strategic focus. As it has a unique capability through its innovative value investment concept, competing through differentiation is an obvious choice. The designing step thus leads to focusing on the three products and offerings characteristics shown in Fig. 7.2 (bottom lightweight business model excerpt). The firm should focus on differentiation through investment concepts rather than solely through new asset class exposures to take a leap step ahead of industry trends. In addition, the firm should compete on absolute return strategies, an industry trend, by leveraging its investment concepts. The focus should be on liquid asset classes allowing to define a clear delineation with the hedge-funds industry.

As can be seen from this example, designing possible prototypes for a firm's strategic focus can be done effectively through abductive reasoning.

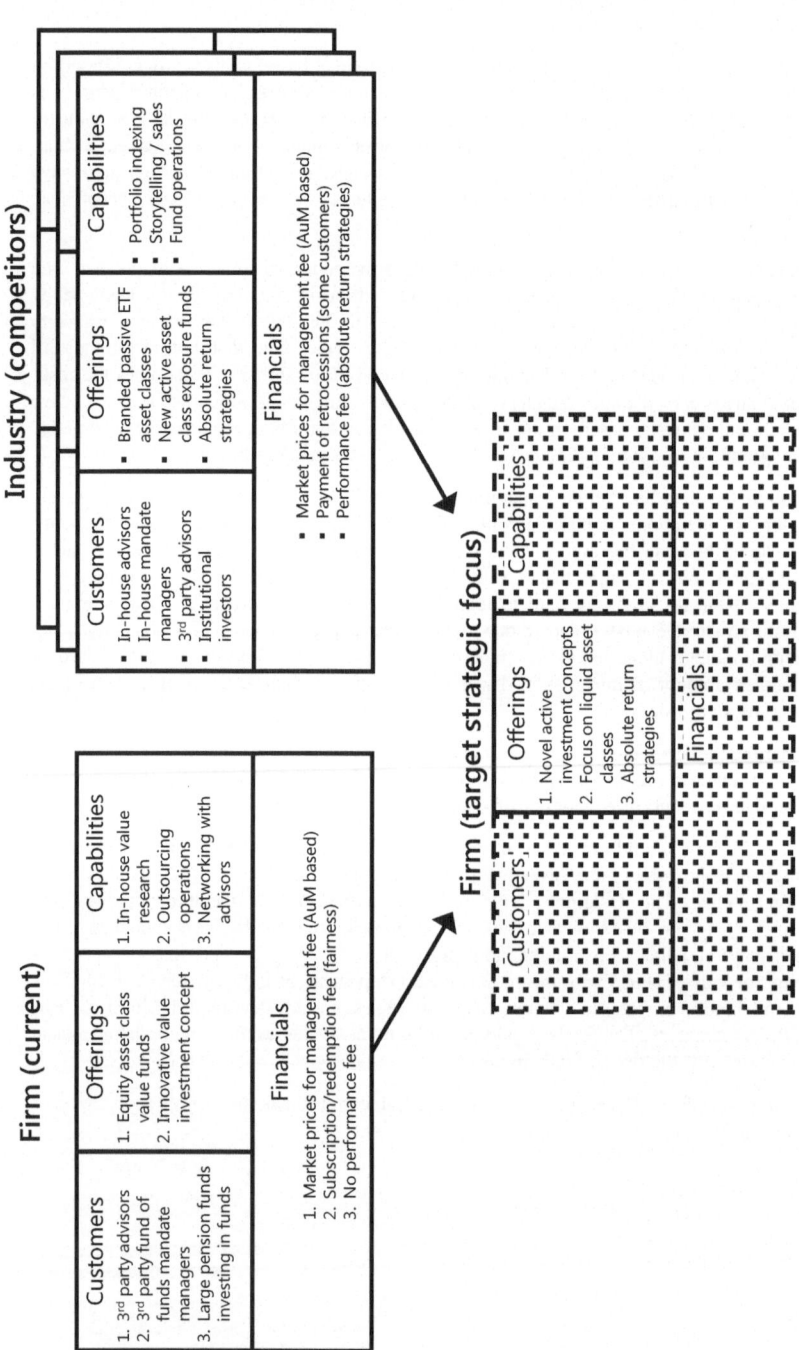

**Fig. 7.2** Example of determining the target strategic focus for an independent boutique fund management firm based on the firm's current strategic focus and the mutual funds industry in Europe (competitors are not shown explicitly)

## 7.4  Validating the Designed Strategic Focuses

Validating the designed strategic focuses is key for success. Validation is based on formulating and testing hypothesis. A *strategy hypothesis* is a testable belief related to future value creation of elements of a strategy (Schrage 2014). A *strategy experiment* is an objective, easily replicable test of a strategy hypothesis that generates measurable insights as to whether the hypothesis is valid or invalid. Neither a strategy hypothesis, nor an associated strategy experiment, validate a strategy in general or the strategic focus in particular, in its entirety.

The goal of validating the prototyped strategic focuses is avoiding failure further down the road of designing the firm's strategy. To those familiar with statistical hypothesis testing (Kuehl 2000), validating strategy hypothesis has some similarities, but many dissimilarities, with hypothesis testing in statistics. Strategy hypothesis testing is not about statistical precision or *t*-values. It is about getting an external first-hand confirmation of internal beliefs on which the strategic focus prototypes are based. Strategy hypothesis testing aims at answering those questions that could potentially change the validity of the prototyped strategic focus.

> **Example** The strategic focus may be defined by the characteristic that customers in the targeted segment of individuals over 65 need mobile phones that have large keys or icons because their visibility is usually poor. Asking a sample of over 65-year-old people to write a 140-character message on an old Blackberry phone with its typical small but ergonomic keyboard in less than one minute, could easily support or invalidate the hypothesis, depending on whether most of the test people were to fail or succeed.

### 7.4.1  Checking for Consistency

Before starting to formulate assumptions underlying a strategic focus prototype, it should be reviewed relative to the external environmental constrains identified during the current environmental analysis. This means, checking each characteristic of the prototyped strategic focus for whether or not it contradicts any existing environmental characteristic. In the case of a contradiction, the strategy design process must revert to the designing step to introduce amendments that fix the violation.

### 7.4.2  Formulating Strategy Hypothesis

When confronted with the task for the first time, formulating strategy hypothesis to test strategic focuses is hard. First, key assumptions are identified by comparing the prototyped strategic focuses to the findings from the customers as well as target industry environment analysis. Differences provide a good starting point for identifying made assumptions.

Consider an industry mainly relying on a financials strategic focus, aiming at providing cheap commodity products. If the designed prototype suggests an offering based strategic focus, a possible assumption would be "the firm is capable of designing, producing, and selling products that are sufficiently different from those of competitors such that customers are willing to pay a premium price." To identify additional assumptions made, the five whys[4] method may be applied. Ask and answer five times the question why, regarding a given characteristic of the strategic focus. The last answer received is often a sound formulation of the hypothesis to be validated.

**Example** Consider a customers strategic focus, aiming at providing mobile phones tailored to people over 65. The five why questions asked and answered could be:

(1) Why does the strategic focus target people over 65? Because they are retired and have more spare time to use their mobile phones.
(2) Why do retired people with significant spare time need tailored mobile phones? Because they are less stressed than people still in active live and as such have different needs.
(3) Why does being less stressed lead to different needs with respect to mobile phones? Because with more spare time available, they have more time to call friends and relatives. In addition, calling friends and relatives makes them less lonely.
(4) Why does calling friends and relatives require different mobile phone features? Because mobile phones aimed at the working population focus their features on all but calling.
(5) Why is using the calling feature on current mobile phones not satisfy the needs of the targeted elderly population? Because elderly people have a hard time learning new technologies and navigating a large set of unnecessary functions.

Out of the five why analysis, one strategy hypothesis to be tested is "elderly people require mobile phones that are easy to use for placing and receiving phone calls."

When formulating strategy hypothesis, the focus should be on those characteristics of the strategic focus that would be invalidated by a failed test. Just because believing that something is valid, does not make it validated. It is common, although not always the case, that strategy designers believe that their formulated hypotheses are true, especially inexperienced strategists. It is important not to fall into the trap to assume that individual beliefs represent the truth and do not require validation.

Any good strategy hypothesis has three characteristics:

(1) The hypothesis *relates to the strategic focus* and its characteristics in such a way that its invalidity would require an adjustment to the strategic focus or make the strategic focus fail altogether.
(2) The hypothesis is *easy to understand* by people knowledgeable with the target industry without having been involved in the strategic focus prototyping activity.

---

[4]The "five whys" method is an iterative technique used to explore the cause-and-effect relationships underlying a particular statement. The "why" question is repeated five times. Each answer forms the basis of the next why question.

(3) An experiment to validate the hypothesis *can be performed quickly*, usually in less than five weeks, *and cheaply*, usually for less than $5000, and with less than five strategy designers being involved.

The hypothesis to be tested to validate a given strategic focus prototype can be identified as special cases of three generic assumptions:

(1) The lightweight business model dimension underlying the strategic focus is sound in the targeted industry based on the environmental analysis.
(2) Competing through differentiation, respectively superiority, is a sound decision.
(3) The designed characteristics of the strategic focus are valid.

### 7.4.3  Designing Strategy Experiments

Designing strategy experiments to validate assumptions requires creativity and a good understanding of the target industry. Any strategy experiment must include the following six characteristics:

(1) The hypothesis to be tested is formulated in a *clear and easy to understand* way.
(2) The experiment describes the *activities to be performed* to test the hypothesis.
(3) The experiment includes a *metric used to measure* the success, respectively failure.
(4) The population as well as the minimal and target *sample size* to conduct the experiment on are defined.
(5) Success as well as failure *criteria* are defined in relation to the measured metric and the sample size.
(6) The *time horizon* as well as the expected *costs* for performing the experiment are identified.

It is not uncommon to start with a relatively small target population sample size, only to increase it when the initial results of the experiment are inconclusive, relative to the success and failure criteria defined. Asking how many additional responses would be needed to change a preliminary result, gives a good indicator of the target sample size. Strategy experiment design should follow the described design thinking principles, that is, focus on individuals, avoid non-value-adding analysis, and use iterations to improve the quality of the results over the course of the performed validation experiments. Validations, that is, the outcomes of strategy experiments, are decision support tools. They should be considered as such and not as a method for confirming an unknown ultimate truth.

**Example** Consider the offerings strategic focus prototype of a small independent fund management boutique as described in Fig. 7.2. A key assumption made is that an offering focusing on a novel value based active investment concept is desirable. This assumption is central to the chosen strategic focus, as well as the decision to compete based on differentiation. To ensure that the assumption fulfills the three characteristics underlying any good strategy hypothesis, it can be reformulated as "given a comparable performance and risk management track record, investors prefer to invest in an innovative value based actively managed mutual fund over more traditional actively managed funds and passive value-based exchange traded funds."

One way of testing that assumption would be to develop a hypothetical KID[5] for the offering. Then, during a fund fair, the hypothetical KID as well as actual KIDs of competing offerings would be made available to potential investors. The number of investors interested in either offerings, measured by the number of KIDs distributed, could be used as metric. The target size underlying the experiment would be the number of visitors during the fair, requiring at least 100 interested visitors. To measure success versus failure, a more than $2/3$ versus less than $1/3$ ratio could be used, defined as the interests of investors in the novel offering versus the traditional ones. The cost of such an experiment is related to the production of the hypothetical KID as well as the presence at the fund fair.

The goal is not to get a statistically significant result, but to gain enough insight that it would be highly unlikely that additional information could change the validity of the chosen strategic focus. As can be seen from this example, it is possible, with a simple and effective process to identify and test a strategic focus of a firm and jump-start the strategy design process without lengthy and unproductive analyses.

## 7.5  Selecting the Target Strategic Focus

Once one or more strategic focus prototypes have been designed and successfully validated, it is time to select the one that is the most appropriate to base the firm's future strategy on. This choice is a key strategic decision and should be made by the decision takers responsible for the firm's strategy. Decision takers should ideally have been actively involved in the design and especially the validation process.

There exist two complementary approaches for choosing the strategic focus. In the first approach, multiple strategic focus prototypes, based on the same lightweight business model component and approach for competing, are merged into a single broader strategic focus. This is the preferred approach if the resulting characteristics are not contradicting or diluting the strategic focus. Alternatively, one of the multiple strategic focus prototypes is chosen on its merits and the other ones are put on hold, to be ready for use if and when the initial choice made is found to be inappropriate during the business model and/or competition layers of the strategy design process. It is explicitly not recommended to continue the strategy design process with multiple strategic focus instances, as this leads to significant irrelevant analysis, design, and validation activities. Rather than being a decision by a single decision taker, or the outcome of a vote, the choice of a strategic focus should result from consensus building among decision takers. This is

---

[5]The KID is a standardized Key Information Document required by the MiFID (Markets in Financial Instruments Directive) directive of the European Union to be provided to any investor ahead of their investment decision.

important to ensure that the decision, and as such, the derived strategy, has a broad backing at the highest level of the organization.

## References

Christensen, C. M., Hall, T., Dillon, K., & Duncan, D. S. (2016). *Competing against luck: The story of innovation and customer choice*. New York, NY: HarperCollins Publishers.

Collins, J. C., & Porras, J. I. (1996). Building your company's vision. *Harvard Business Review, 74*(5), 65–77.

Cross, N. (2011). *Design thinking*. London, UK: Bloomsbury Academy.

Diderich, C. (2017). Initiating the strategy process using design thinking. *Change Management Strategy eJournal, 9*(8). SSRN: https://ssrn.com/abstract=2927941 or http://dx.doi.org/10.2139/ssrn.2927941.

Grant, R. M. (1991). The resource-based theory of competitive advantage: Implications for strategy. *California Management Review, 33*(3), 114–135.

Harris, J. D., & Lenox, M. J. (2013). *The strategist's toolkit*. Charlottesville, VA: Darden Business Publishing.

Kuehl, R. O. (2000). *Design of experiments: Statistical principles of research design and analysis*. Boston, MA: Duxbury-Thomson Learning.

Porter, M. E. (1985). *Competitive advantage*. New York, NY: The Free Press.

Schrage, M. (2014). *The innovator's hypothesis*. Cambridge, MA: MIT Press.

Treacy, M., & Wiersema, F. (1995). *The discipline of market leaders: Choose your customers, narrow your focus, dominate your market*. New York, NY: Perseus Books.

Utterback, J. M. (1994). *Mastering the dynamics of innovation*. Boston, MA: Harvard Business Review Press.

Zott, C., & Amit, R. (2013). The business model: A theoretically anchored robust construct for strategic analysis. *Strategic Organization, 11*(4), 403–411.

Zott, C., Amit, R., & Massa, L. (2011). The business model: Recent developments and future research. *Journal of Management, 37*(4), 1019–1042.

# Part IV
# Iteratively Developing the Business Model Underlying the Strategy

# Gaining Insights by Observing Target Customers in Their Natural Environment

> *Learn from yesterday, live for today, hope for tomorrow.*
> *The important thing is not to stop questioning.*—Albert Einstein

During the foundation layer, the *target industry* in which to compete has been identified and the field of play, the so-called *strategic focus*, selected. The second layer of the strategy design process, the business model layer, focuses on designing how the firm wants to create value for its customers, and subsequently, its stakeholders. It can be decomposed into four process steps, the four steps of design thinking, that is, observing, learning, designing, and validating. This chapter focuses on the first step, exploring target populations through observing. The *target population* is defined as subjects, such as customers, employees, or suppliers, or objects, such as technologies, capabilities, or processes, that are important relative to the strategic focus and are at the center of the firm's business model. The term target population is a generalization of the term target customer. Depending on the chosen strategic focus, observing means laying the foundation to understand customer jobs-to-be-done, identifying innovation expertise, detecting capabilities and resources, or apprehending financial traits. During the observing step, a firm-specific perspective is taken, such as focusing on customers, suppliers, and employees, rather an industry specific perspective, such as focusing on competitors.

## 8.1 Observing Objectives

The goal of the observing step is to identify insights that may be of value during the designing step. Stated otherwise, the focus is on laying the groundwork for subsequently generating innovative ideas onto which the firm's target business model and strategy can be based. To do so, the target population is observed in its natural environment, looking for analogies, associations, and contradictions. A special case of observing is listening (Verganti 2009). Depending on whether the firm is a

© Springer Nature Switzerland AG 2020
C. Diderich, *Design Thinking for Strategy*, Management for Professionals,
https://doi.org/10.1007/978-3-030-25875-7_8

mature firm, operating in a well-defined environment, a start-up, following a greenfield approach, or a firm seeking to disrupt its current environment, the sought-after insights will be different. To avoid being distracted and wasting resources, the observing step focuses primarily on those elements of the business model that are related to the strategic focus. Although the observing step exhibits similarities with a traditional SWOT analysis, the focus is on observing rather than interpreting. It is important not to start designing the target business model or look for insights to confirm a pre-sought solution.

### 8.1.1 Observing Mature Firms

The outcome from the observing step in the context of mature firms includes observations from

- the firm's *current detailed business model* elements related to the chosen strategic focus and their relations with the value proposition (OVP) and the products and services (OPS) elements,
- *declining and failing elements* of the existing detailed business model that have an impact on the chosen strategic focus, as well as served customer segments, their jobs-to-be-done, and their willingness to pay, and
- the firm's *strengths* and *weaknesses* in any of the 15 elements of the firm's current detailed business model with an impact on the strategic focus.

If a customers strategic focus is chosen, customers and their jobs-to-be-done are at the center of the observing step. Existing capabilities are less relevant to observing, and thus need less attention, as they will anyhow be re-defined at a later stage based on the customer jobs-to-be-done chosen to be satisfied by the offered value propositions.

### 8.1.2 Observing Start-up Firms

Start-up firms or firms aiming at entering an emerging industry should focus on observing potential unmet customer needs. They must look for failures in business models of potential competitors and aim at identifying areas that could be improved or exploited.

**Example** A typical example is a robo-advice strategy implemented by a start-up wealth management firm. During the observing step, it is identified that customers are becoming more and more price sensitive, especially after numerous failed attempts by traditional wealth and asset management firms to deliver value, that is, investment performance based on skills. In addition, inefficiencies in business processes underlying the implementation of portfolio changes are identified.

If following an offerings strategic focus, emerging technologies are at the center of the observing step.

| Customer segments | Value proposition | Capabilities |
|---|---|---|
| – Stock exchanges<br>– Luxury goods dealers<br>– Pharmaceutical R&D<br>– Individuals in countries with unstable regimes<br>– Financial auditors<br>– Private markets<br>– Software providers | – Real-time settlement<br>– Authenticity and traceability of ownership<br>– Immutability of clinical trial results<br>– Independence from government<br>– Simplification of audit processes<br>– Peer-to-peer transactions, avoiding trusted intermediaries<br>– Addressing licensing issues | – Public key cryptography<br>– Consensus building algorithms<br>– Randomization<br>– Decentralized database replication<br>– Algorithmic trust |

**Fig. 8.1** Sample subset of insights gained by observing the implementations and the use of existing blockchain technology-based solutions

> **Example** A start-up aiming at competing in the general ledger industry implementing an offerings strategic focus, may study different applications of blockchain technology. Figure 8.1 illustrates some of the insights gained from observing existing blockchain solutions, classified along the customer segments, value proposition, and capabilities elements.

## 8.1.3   Observing Disruptors

Understanding the potential for disruptor firms requires identifying on one side, unmet customer needs or needs that are only met in an unsatisfactory way, and, on the other side, approaches or technologies that have not yet been related to the identified needs.

> **Example** Airbnb identified the need for renting individual rooms to strangers, both from a supplier and a customer perspective, on the one side, and the need for an internet platform technology matching supply with demand in a very efficient and cost-effective way, on the other side.

> **Example** Figure 8.2 illustrates the outcome of observing the automotive industry with a focus on disrupting.

To avoid focusing only on those pairs of needs-approaches that exhibit a potential match, the two sides, customer needs and approaches/technologies should

| End-customer jobs-to-be-done | Approaches/Technologies |
|---|---|
| – Reduce up-front investments | – Battery technology |
| – Contribute to clean environment | – Sharing platforms |
| – Ensure high autonomy | – Artificial intelligence |
| – Gain freedom/independence from 3$^{rd}$ parties | – Leasing models |
| – Focus on safety first | – Airbag technology |
| – Automate driving and parking | – Stringent security regulations |

**Fig. 8.2** Sample outcome from an observing analysis focusing on disrupting the automotive industry

be observed by distinct strategy sub-teams. It is important to note that observations may or may not be relevant for subsequent ideation and designing decisions. They must be objective, specific, and concise.

## 8.2  Deriving Perspectives Based on the Strategic Focus

The aim during the observing step differs depending on the underlying strategic focus and the target populations considered.

*Customers strategic focus* The focus should be along the two dimensions:

(1)  Identifying, characterizing, and clustering customers.
(2)  Understanding customer needs, both met and unmet, and focusing on their jobs-to-be-done.

The customers and their felt pains and sought-after gains need to be at the center of the observing step.

*Offerings strategic focus* When following an offerings strategic focus, observing must target insights, such as identifying new technologies, that may be used to create new customer needs or address existing customer needs in a novel way. The challenge is on gaining insights by observing the present that may be used to change the future. Innovation often results from re-configuring existing insights and knowledge in different ways. As such, the observing step must focus on identifying insights that may lead to knowledge which can be re-configured and re-combined to innovate.

*Capabilities strategic focus* Observing capabilities of a firm aims at understanding what a firm is good at. Capabilities may be around executing business processes, owning specific knowledge and intellectual property, using unique technologies and tools, or possessing unique access to capital and investors.

*Financials strategic focus* The focus during the observing step, when focusing on financials, is on how different stakeholders, customers, suppliers, investors, to name the most important ones, perceive money (pricing models) and the flow of money over time (cash flows).

## 8.3   The Observing Process

Observing the target population to gain new insights is one of the four core activities in design thinking. The observing process O is based on an ethnographic approach (Spradley 1979, 1980; Liedtka et al. 2014). It is executed multiple times in an iterative way, based on the different target populations identified. Each of the observing steps O.2–O.4 is followed by a learning step L to derive knowledge from the observed insights (as described in Chap. 9) and lay the foundation for the next observing step. Iteratively observing and learning is key to ensure that the design thinking methodology focuses on those aspects that matter most, avoiding non-value-adding activities.

**Process O—Observing**

O.1   Identifying target populations related to the chosen strategic focus
O.2   Passively observing informants in the target populations
L      *Learning from observed insights (see Chap. 9 for details)*
O.3   Conducting ethnographic interviews with informants, elaborating on the observations from step O.2
L      *Learning from interview insights (see Chap. 9 for details)*
O.4   Running focus groups to extend the outcome from passively observing (O.2) and ethnographic interviews (O.3)
L      *Learning from focus group insights (see Chap. 9 for details)*
O.5   Using secondary research to gain a different perspective on previously identified insights
L      *Learning from secondary research insights (see Chap. 9 for details)*

## 8.4   Identifying Target Populations

Observing starts by identifying who or what to observe, the so-called *target populations*. Individuals in the target populations are called *informants*. The personas framework is best suited for defining target populations to observe (So and Joo 2017). It allows setting a reference point using an easy to understand language. As design thinking for strategy allows for non-customer-centric design activities, the target populations, as well as the informants, may be subjects (people, groups, etc.)

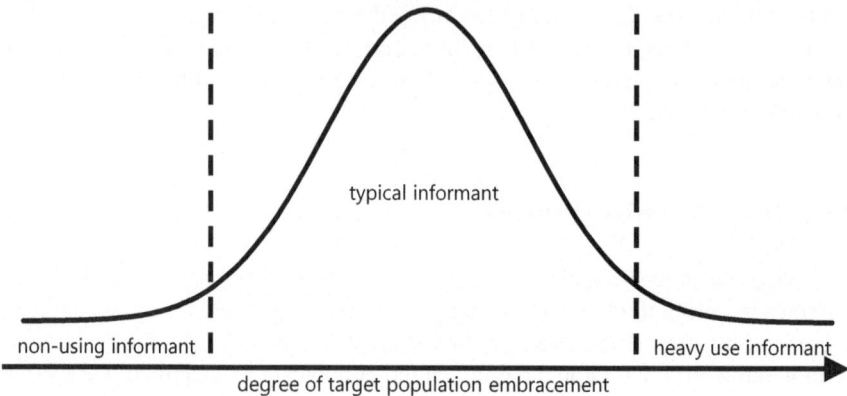

typical informant

non-using informant |     | heavy use informant

degree of target population embracement

**Fig. 8.3** Degree of sophistication of the target population, allowing to identify extreme informants (informants may be subjects, like customers, or objects, like technologies)

or objects (technologies, capabilities, etc.). When choosing informants in the target population to observe or interview, it is important not only to focus on typical informants. Significant insights can often be gained by observing extreme informants, as illustrated in Fig. 8.3, as well as lead users[1]. Indeed, those extreme informants are often able to offer more distinct input as they have thought about the issue at hand more thoroughly beforehand.

> **Example** Consider early adopters of the 5G mobile communication technology using one of the first mobile phones offering 5G data access. They can provide significant insights with respect to the network speed and its relevance to the applications they use mobile data for, for example when the new technology is perceived as disruptive versus when it only offers marginal advantages. These insights may lead to unique areas of deployment of the 5G technology and formulating a distinct business model for a dedicated target population.

The personas framework supports defining the target population. It has been developed and is predominantly used to describe subjects, especially customer segments, to target in a personalized way. It can also be applied to describe objects, that is, technologies, capabilities, or activities. A *persona* is defined by four key characteristics:

(1) A *name*, giving a humanized description to the persona.
(2) *Screening questions*, allowing to segregate whether a subject or object belongs to the persona or not.
(3) A *description*, telling a story on what the persona is related to and why it is important.

---

[1] A lead user is a customer of an offering whose value to a broader audience is still unknown. Lead users are typically early adopters helping define the value of an offering from the customer perspective, before there actually exists a marketplace for it.

| Persona name | Jennie, the executive assistant | John, the in-house travel agent |
|---|---|---|
| Screening questions | – Is assistant to an executive?<br>– Organizes travels?<br>– Is responsible for the bookings made? | – Is part of a central unit organizing travels?<br>– Is responsible for bookings made?<br>– Does not report to the executives who are traveling? |
| Description | Assistant supporting one or more executives with their time management activities, including coordination and organization of business travels | Travel agent focusing on specific needs of in-house executives traveling, aiming at optimizing travel costs |
| Characteristics | Time-optimized travel | Cost-efficient travel |

**Fig. 8.4** Sample persona definitions focusing on organizing business travelers

(4) Underlying *characteristics*, relevant in the context of the target industry and the strategic focus, for example, common jobs-to-be-done, similar algorithms, or shared skills.

Personas are especially relevant when their characteristics can be related to the real world, to properties that can be observed. Good personas are testable for their existence and relevance. Demographic descriptions are not considered good personas as they do not offer good characteristics related to the target industry and the strategic focus.

**Example** Figure 8.4 illustrates two different personas that can be used to observe online travel reservations in a business environment, aiming at designing a customers strategic focus-based strategy.

Although customers may be included in any target population, their personas significantly differ based on the target strategic focus. Depending on the strategic focus chosen, different target populations are in focus.

## 8.4.1  Customers Strategic Focus-Based Target Populations

When following a customers strategic focus, the target populations are defined by current and potential customers. Historically, customer focused target populations are defined based on geographic, such as living location, family residence, or place of work, and demographic, such as race, ethnicity, age, gender, religious, education,

**Table 8.1** Sample set of generic definitions of target populations based on different strategic focuses

| Customers strategic focus | Offerings strategic focus | Capabilities strategic focus | Financials strategic focus |
|---|---|---|---|
| – Students<br>– Single households<br>– Families with children<br>– Military personnel<br>– Retirees | – Blockchain<br>– Cryptography<br>– Battery technology<br>– Connecting platforms<br>– Social media | – Process optimization<br>– Supply chain management<br>– Real-time software development<br>– Planning and scheduling | – Cheapest price<br>– Value based pricing<br>– Subscription based<br>– Competitive pricing<br>– Cost optimization |
| (a) | (b) | (c) | (d) |

income, marital status, or occupation, characteristics. More recently, *psychographic characteristics*, such as lifestyle, values, social class, and personality, as well as *behavioral patterns*, such as usage, loyalties, awareness, occasions, knowledge, liking, and purchase patterns, have become popular in defining customer centric target populations. Table 8.1 a illustrates typical customer-based target populations. Once identified, these target populations are refined, and specific personas are associated to them.

### 8.4.2   Offerings Strategic Focus-Based Target Populations

Innovation and expertise stand at the forefront when defining the target populations in an offerings-based strategy. Often customers targeted with new innovative offerings are considered being the target populations. Better insights are gained when defining inventions, new technologies, or disruptive processes, as the target populations to consider. For example, the blockchain technology may be used as a target population. Or, in a more generic setting, cryptographic technologies, such as public key encryption or hashing algorithms, may be used as the target populations for observing. Table 8.1b illustrates typical offerings focused target populations.

### 8.4.3   Capabilities Strategic Focus-Based Target Populations

When focusing on leveraging capabilities as the core superiority characteristic, identifying key process capabilities is important. For example, supply chain management or real-time software development, may be used as target populations to observe, as illustrated in Table 8.1c. Looking at the value chain of the firm or the industry allows identifying possible capabilities-based target populations.

### 8.4.4 Financials Strategic Focus-Based Target Populations

Identifying target populations in a financials strategic focus environment requires describing different personas that have an active impact on revenues and/or costs, either in terms of value or structure. Typical target populations are defined around economic buyers, decision makers, suppliers, and internal cost owners. Table 8.1d illustrates possible financials-based target populations. Alternatively, target populations may be defined around pricing models, for example, those populations that prefer paying a lump sum, those that feel at ease paying on a time and material basis, or those that look for volume-based fees.

---

It is important to note that there does not exist a single best set of target populations to consider. The aggregation of all target populations considered should be broad and cover at least 80% of all potential populations of interest. Identifying a target population does not mean that the firm wants or has to serve that target population, but merely that it is relevant in the target industry. Defining target populations and associated personas is as much an art as it is a science. Strategy designers should adjust the definitions of target populations and personas used throughout the observing and learning steps, depending on new insights and knowledge gained.

---

## 8.5 Passively Observing

Once the target populations have been identified, the passively observing step (L.2) starts. It is called passively observing because the goal is not to interfere or interact with the informants observed. It is through that non-intrusion that actual challenges can be best observed, and insights identified. This is in stark contrast with traditional analytical approaches that start with questioning the informants. Passively observing is related to gaining insights that can serve as the basis for ideation, rather than collect statistically significant data. It aims at going at least one layer below the surface and identifying not obvious insights.

> **Example** One may observe an executive assistant when she makes a travel reservation online. One of the insights gained may be that, just before confirming the to be made reservation, she reviews it to ensure that it is compliant with the firm's travel policy. This observation related to compliance may lead to identifying a pain point during the learning step and, address it in the value proposition development, during the designing step. If starting with interviewing informants, such an insight may be overlooked as the informant sees this as obvious and forgets to mention it.

## 8.5.1  Types of Observations

Ethnographic observation approaches are at the core of passive observing. They can be classified into three main categories, that is, grand-tours, mini-tours, and in-depth observations, depending on the breadth and depth of the targeted observations (Spradley 1980). Passive observation always starts with a grand-tour.

*Grand-tour*  The grand-tour observation approach provides a high-level holistic view of the target population. It focuses on understanding the big picture and identifying areas where mini-tours are of value. Typically, 20% of the time spent on observing a target population is spent on an initial grand-tour observation.

> **Example** A typical grand-tour observation focusing on grocery store customers as target population aims at identifying the different activities a customer performs in a grocery store, from selecting a shopping trolley, choosing vegetables, asking for advice from the butcher, to paying at the cashier.

*Mini-tour*  In ethnography, the focus areas of a mini-tour is called a *domain*. During each mini-tour, one or more specific domains identified during the grand-tour are investigated. Two to five mini-tours, taking up about 50% of the overall time spent on observing, support gathering insights related to specific topics identified during the grand-tour. Identifying the right domains that require mini-tour observations is a key skill strategy professionals need to exhibit.

> **Example** Mini-tours related to observing grocery store customers could be around the activities supporting choosing vegetables, looking for advice from the butcher, or the overall payment and check-out process, including packing the bought goods.

*In-depth observation*  The remaining 30% of the time spent on observations should be dedicated to in-depth observation, aiming at understanding the specificities of insights gained during mini-tour domain observations. Typically, each mini-tour observation leads to one to three in-depth subsequent observation sessions.

> **Example** During an in-depth observation session, following a mini-tour on the payment and check-out process in a grocery store, the domains observed could be focusing on the different payment methods used, like paying with cash, using credit or debit cards, or even reverting to mobile payment solutions, such as ApplePay or SamsungPay.

The identification of mini-tour and in-depth observation domains to investigate is a dynamic and iterative process. It is not possible to define beforehand in detail all aspects that need to be observed. Ethnographic observers always look for hints from the informants where to focus the next observation activities. Figure 8.5 illustrates the relationship between the different types of observations with respect to the breadth and depth of insights gained.

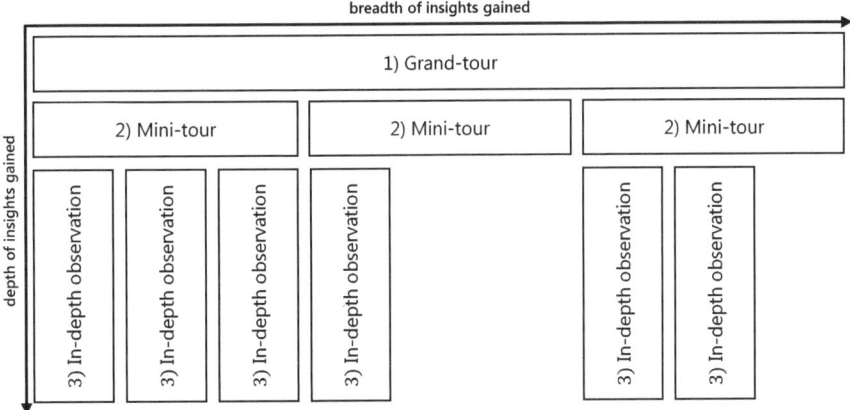

**Fig. 8.5** Selecting the order and scope (breadth versus depth) for the various types of passive observation steps

## 8.5.2 Passively Observing Process

The passively observing process can be decomposed into three iterative steps, the observation preparation step (O.2.1), the informant observation step (O.2.2), and the recording and documentation step (O.2.3). The outcome of passive observations is a list of raw, not interpreted, insights, and a set of unaddressed questions. Successful passive observation is as much an art as it is a science. A key skill that design thinking-based strategy professionals need to learn is identifying what may be relevant to observe and document, and what can be safely ignored. Answers to the following two questions help address that challenge:

(1)  Are the to be observed insights relevant from the informant's perspective?
(2)  Are the to be observed insights important to know from the observer's perspective?

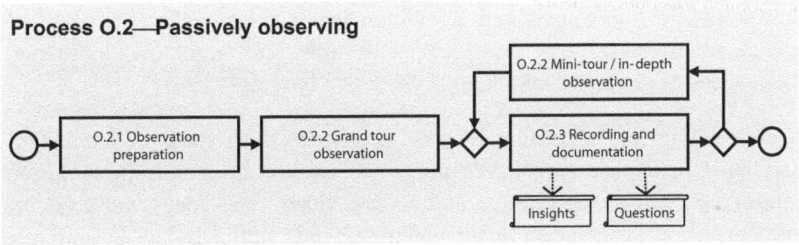

Successful observing needs observing others, not oneself. It is easy to fall into the trap assuming that the way one does things is identical to the way others would do the same things. Questions asked during passive observation should be kept to a minimum and only focus on understanding, not interpreting. During the observing step, why questions should never be asked. They would lead to distracting the informant and potentially bias the observed insights.

The biggest challenge faced during passively observing is avoiding information overload, that is, focusing on irrelevant details, without ignoring those details what would traditionally be ignored or blocked out. Observing requires increasing awareness. Successful observing focuses on empathy keeping

– the eyes wide open,
– the ears in listening only model, and
– the mouth shut.

### 8.5.3  Passive Observation Tools

The most prominent tool used to support passively observing is the *thinking aloud* tool. It requires the informants to verbally describe what they think during the different activities that they perform.

---

**Tool—Thinking Aloud**

Thinking aloud is a tool that supports the observation process by making the thoughts behind the observed activities transparent. While performing the activities, the informant describes their thoughts that underlie them. The observer is documenting the spoken protocol jointly with the observed activities. The spoken worlds may describe what the informant is looking at, what he is thinking, doing, or feeling.

A typical thinking aloud application could be documented as follows. While booking an airline travel online, the executive assistant says that she is first screening the flights based on the target arrival time, and only thereafter checking if the prices offered are within the travel guidelines of the firm, which she reviews by accessing a specific web page on the firm's intranet.

---

Documenting and recording insights should be done hand in hand with the observing step. Typical tools to document observations are notes, drawings, and, if the informant allows it, photos, audio, and video. The focus must be on the informant. Good observation documentation avoids summarizing and interpreting the insights gained. This is done at a later stage, during the learning step, described in Chap. 9. While documenting observations, any open question related to what is missing, is written down. These questions, either support iterating the passively observing step (O.2), adding additional sub-tours, in-depth observations, or serve as

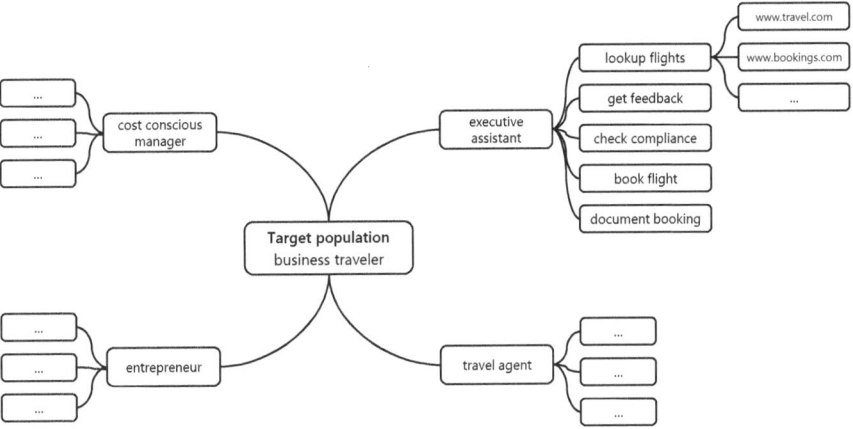

**Fig. 8.6** Mind map documenting observations from informants ordering airline tickets on-line, focusing on business travelers

the basis for preparing and conducting ethnographic interviews (O.3). Typical tools used for recording observations are *mind maps* and *storyboards*. Figure 8.6 illustrates a mind map derived from observing a grand-tour of four different informants ordering airline tickets on-line.

### Tool—Mind Mapping

A mind map is a two-dimensional diagram visually organizing multiple insights in a hierarchical way. At the center of the diagram is the core topic covered, a target population, a persona, or a domain, usually identified ahead of or during a grand-tour or a mini-tour observation round. Branches describe characteristics of the core topic observed. Each branch may have zero or more sub-branches that describe observations related to the sub-topics. Each branch may describe a sub-domain and document related insights from mini-tour or in-depth observation rounds.

### Tool—Storyboard

A storyboard is a sorted collection of graphical illustrations or photos documenting a sequence of observed insights. Storyboards help documenting the timeline behind observations. Each element of a storyboard describes a separate activity observed. Using illustrations or photos rather than text is based on the insight that an image says more than a thousand words. Illustrations are easier to understand and communicate than descriptive text.

## 8.6  Conducting Ethnographic Interviews

Only observing informants does not tell the whole story or give all the insights needed to design a successful business model underlying a strategy. Ethnographic interviews (Spradley 1979; Liedtka et al. 2014) take up the process where passive observation left off. They focus on better understanding the insights gained from informant observations. Conducting ethnographic interviews can be subdivided into three activities, that is, preparing, conducting, and documenting the interviews. In addition to interviewing informants, it is best practice to also conduct interviews with interpreters (Verganti 2009).

> ### Process O.3—Ethnographic Interviews
>
> O.3.1  Preparing ethnographic interviews (selecting the focus area, identifying open questions)
> O.3.2  Conducting the interviews (keeping the informant speaking 80% of the time)
> O.3.3  Documenting the insights gained

Ethnographic interviews are only as good as their preparation and the questions asked. Good interviewing preparation starts with selecting a focus area. The focus area is chosen based on the outcome of the passive observation step (O.2) and the identified unanswered questions. Interview questions are formulated as open questions that support the informant talking. Typically, ethnographic interviews include three types of questions:

(1) Descriptive questions of the form "could you describe", "could you tell me", or "what do you do"?
(2) Structured questions, focusing on giving structure to the answer from descriptive questions, such as "what are the possibilities" or "what are meaningful alternatives"?
(3) Contrast questions, aiming at understanding the meaning of the answers received so far, such as "what is the difference between" or "why is it that way and not another way"?

While conducting the interviews, it is important to remain open-minded, but avert the interview getting off track. Ethnographic interviews start with descriptive questions avoiding formulating pre-assumed answers into the questions. Conducting good interviews, that is, interviews providing a lot of insights is hard. Inexperienced interviewers should test run their interviews before addressing real informants. The saying "your never get a second chance to make a first impression" applies. Good interviewers express interest in the informant and show ignorance. They look beyond the obvious and make the informant rather than the interviewer feel to be the expert.

A successful tool used in ethnographic interviews, especially when framing contrast questions, is the *five why* tool. It is an interactive interviewing approach used to explore cause and effects. Each answer forms the basis of the next question.

### Tool—Five Why

The *five why* tool is an iterative questioning technique based on subsequently asking five times the question *why*. Each answer forms the basis for the next why question. This allows getting to the foundation behind the initial question asked.

**Example** A typical five why application, focusing on understanding how an executive assistant executes their online travel booking process, could be documented as follows:

(1) Why are you using the web site www.travel.com? Because it includes all major airlines.
(2) Why is it important to include all major airlines during online travel booking? Because it allows finding a flight that minimizes the waiting time between the arrival time and the meeting time.
(3) And why is it important to minimize the waiting time? Because the executives do not like to wait.
(4) Why do they not like to wait? Because they want to use their time as productively as possible.
(5) And why is it important to use time as productively as possible? Because they need the time to achieve their goals and a day has only so much hours.

The gained insights lead to identifying the activity of prioritizing the use of time of the manager as a key characteristic, something that would not have been obvious by merely observing the executive assistant using a given web site to make a reservation.

As with passive observation, it is important to document the insights gained during, or, at least, at the end of each interview. While doing so, the answers obtained should be related to observations made during the passive observation step. Interviews allow putting the observed insights into perspective. They give an individual viewpoint of the target population observed. Looking for differences between what informants say and what they do is important.

As with passively observing, ethnographic interviews should be followed by a learning step, described in Chap. 9. It is sound to iterate back to the passively observing step to conduct additional mini-tour or in-depth observations based on specific insights from the interviews.

## 8.7   Running Focus Groups

Focus groups, the third approach to gaining insights, aim at extending the knowledge from passive observation and ethnographic interviews, by giving it a group perspective. The goal is to consolidate insights gained and flatten out or understand discrepancies identified. It aims at questioning the validity of individual insights.

Focus groups should be run under the *Chatham House Rule*[2]. The process of running focus groups to gain new and extend existing insights can be decomposed into four steps.

---

**Process O.4—Focus Groups**

O.4.1   Planning questions
O.4.2   Identifying a neutral moderator and selecting informants
O.4.3   Moderating the focus group
O.4.4   Documenting insights gained

---

As with ethnographic interviews, the success of focus groups depends on the preparation and questions asked. This means knowing what one wants to learn. Questions should be open-ended and neutral. They should focus on addressing discrepancies identified between and among observations and interviews. The goal is to allow consolidating the insights gained. Focus groups provide an ideal platform for gaining a group, rather than an individual, perspective on the topics at hand.

The involved individuals in focus groups can be subdivided into three categories:

(1) The *moderator*, ideally a person external to the strategy team or even the firm, with strong moderation skills, but limited subject matter expertise.
(2) The *informants*, a diverse group of six to eight individuals per focus group, different from those observed or interviewed informants, but relating to the topic at hand.
(3) The *analysts*, a small group of strategy team members responsible for identifying and documenting insight from the focus group participants.

---

[2]When a meeting is held under the Chatham House Rule, informants are free to use the information received, but neither the identity nor the affiliation of the speaker(s), nor that of any other informant, may be revealed. The rules originated in 1927 from the headquarters of the UK Royal Institute of International Affairs, the Chatham House.

The role of the moderator is to provide a comfortable environment in which the informants are empowered to share insights. As with interviewers, moderators need to be open-minded, but neutral. They need to ensure that all informants have a say and that different opinions are permitted and encouraged. Questions should be asked one at a time and the moderator must show empathy towards the informants.

Informants should come from a diverse background but be aligned with the target population. They should share their stories and put their individual answers into perspective of the group's insights. They do not have to be experts in the field at hand.

The analysts, responsible for documenting the insights gained from the focus group, must not be part of the focus group discussion itself. They must not intervene. They may sit in the second row, or, ideally, observe the focus group informants through a one-way window or a one-way audio-video channel.

At the end of the focus group step (O.4), the strategy team should have significant insights into the target population, both from an individual as well as from a group perspective. These insights, transformed into knowledge by applying the learning process (see Chap. 9), form the basis for the designing step of the strategy design process, that is, ideating and prototyping new and enhanced detailed business models.

Before doing so, one additional exploratory step is recommended, namely putting the gained insights into perspective through reviewing secondary research.

## 8.8 Performing Secondary Research

To avoid biases introduced by the strategy team conducting the passive observations, the ethnographic interviews, and documenting insights from focus groups, the observing step of the strategy design process should be completed with secondary research. The goal of the last observing step (O.5) is to scrutinize the gained insights and put them into perspective. In contrast with traditional approaches, secondary research is performed at the end of the observing step rather than at the beginning, to avoid non-value-added research.

During secondary research, information from third parties are sought-after in order to validate or invalidate the insights gained so far. The goal is to strengthen the confidence into identified insights, rather than identify new insights.

Sources for secondary research are multiple. Independent research reports may be used. External subject matter experts may be interviewed. Participation in trade shows and conferences may be relied upon to interact with and exchange ideas.

## 8.9    Timeline and Required Skills

There does not exist a single best timeline for the observing step. Two characteristics, in addition to the number of iterations performed, influence the timeline.
These are

- the *experience of the strategy team* members in identifying insights through
  passively observing, conducting ethnographic interviews, performing focus
  groups, and doing secondary research, and
- *the number of target populations* identified.

The overall duration of the observing step may last from one day, when considering an offering-based strategic focus, focusing on a single well-identified
innovation, to multiple months when considering multiple target populations in a
customer focused strategy. Table 8.2 illustrates typical units for characterizing the
time required by the observing steps of the strategy design process. When observing
internal capabilities, rather than external personas, a smaller number of informants
may be sufficient.

Successful strategy designers dynamically adjust the timeline on a need to
identify new insights basis.

The observation step requires four different types of skillsets:

(1) *Strategy experience* that permits identifying those insights which may be rel-
    evant and discard those that are potentially irrelevant.
(2) *Interviewing expertise* to conduct ethnographic interviews with the goal of
    maximizing the quantity and quality of consistent insights gained.
(3) *Access to external moderation expertise* to successfully conduct focus groups.
(4) *Traditional business analyst capabilities* to document the findings and conduct
    secondary research.

At the end, determining the timeline and required resources, is a compromise
relative to the number of iterations of the observing and learning steps needed to
achieve sufficient insights that allow a successful execution of the designing step.

**Table 8.2**  Typical units underlying sound observations per target population

| Step | Units (per target population) | Time |
| --- | --- | --- |
| O.2 Observing sessions | 10–20 observation sessions | 1–2 h per observation session |
| O.3 Interviews | 5–10 interviews | 30–60 min per interview |
| O.4 Focus groups | 3–5 focus group sessions | 2–3 h per focus group |
| O.5 Secondary research | | 2–5 days |

# References

Liedtka, J., Ogilvie, T., & Brozenske, R. (2014). *The design for growth field book*. New York, NY: Columbia University Press.

So, C., & Joo, J. (2017). Does a persona improve creativity? *The Design Journal, 20*(4), 459–475.

Spradley, J. P. (1979). *The ethnographic interview*. Long Grove, IL: Waveland Press.

Spradley, J. P. (1980). *Participant observation*. Long Grove, IL: Waveland Press.

Verganti, R. (2009). *Design-driven innovation*. Boston, MA: Harvard Business Press.

# Understanding Target Populations and Their Jobs-to-Be-Done Through Learning

*An investment in knowledge pays the best interest*
—Benjamin Franklin

During the exploratory observing step, a diverse set of information has been collected. Observing is followed by learning, focusing on making sense of this data. Learning is a confirmatory step. It aims at retaining key insights and transforming them into knowledge to be used during the designing step of the strategy design process. Some design thinkers use the term *sense making* instead of learning (Mootee 2013) as the goal is to make sense of what has been observed. Others call the transformation process *interpreting* (Verganti 2009). The objective is to understand the present by creating a mental model or a map that structures the gained insights and transforms them into usable knowledge, focusing on the firm and its relationship with the environment, primarily customers and their jobs-to-be-done.

## 9.1 Learning Objectives

The learning step focuses on retaining, sorting, aggregating, and structuring insights gained from the observing step. Insights are clustered using various generic as well as specific frameworks, to synthesize knowledge. The derived knowledge serves as the basis for designing the firm's future business model and strategy. Learning, consistent with observing, is on gaining knowledge around the strategic focus.

© Springer Nature Switzerland AG 2020
C. Diderich, *Design Thinking for Strategy*, Management for Professionals,
https://doi.org/10.1007/978-3-030-25875-7_9

## 9.2  The Learning Process

The learning process L contains three key activities:

(1) *Choosing a reference point*, a framework.
(2) *Relating the observed insights* to the reference point through model building activities.
(3) *Deriving knowledge* that matters towards the design of the firm's future business model and strategy by interpreting the observed.

Although described as a separate process, learning is tightly intertwined with observing.

---

**Process L—Learning**

L.1  Selecting a framework to be used for mapping the observed insights
L.2  Mapping and clustering observed insights from process O using five model building activities (associating, rephrasing, calibrating, identifying interdependencies, and formulating abstractions), onto the selected framework
L.3  Formulating assumptions and open questions underlying the gained knowledge
L.4  Validating assumptions and answering open questions

---

## 9.3  Selecting a Framework

Learning starts by selecting a framework that provides a structural foundation to classify the observed insights and derive knowledge. Frameworks also provide a common language to all participants. They help identify blind spots to address in upcoming iterations of the observing process O. The frameworks can be classified into two categories, those that focus on insights from grand-tour observations, and those that are better suited for processing mini-tour and/or in-depth observation insights. The former focus on the breath of the insights whereas the latter focus on depth. Table 9.1 provides a non-exhaustive list of frameworks that can be used.

Depending on the context, the used framework may be customized, expanded, or simplified. The goal is not on mapping insights onto a framework, but on deriving knowledge from the mapping using the framework as a support tool.

**Table 9.1**  Sample list of frameworks and their use

| Observation type and focus area | Framework | Reference |
|---|---|---|
| Grand-tour observation | Lightweight business model | Chapter 3, Diderich (2017) |
| | Detailed business model | Chapter 3 |
| | Business model canvas | Osterwalder and Pigneur (2010) |
| | Ethnographic interview | Chapter 8, Spradley (1979), Liedtka et al. (2014) |
| | Kotler's 4P of marketing | van Assen et al. (2009) |
| | Five forces analysis | Porter (1979) |
| Mini-tour focused on customer jobs-to-be-done | Personas | Chapter 8, Liedtka et al. (2014) |
| | Jobs-to-be-done | Christensen et al. (2016a, 2016b), Liedtka et al. (2014) |
| | Customer decision journey | This Chapter, Court et al. (2009) |
| | Journey map | This Chapter, Liedtka et al. (2014) |
| | Value proposition | Osterwalder et al. (2014) |
| Mini-tour focused on capabilities | Value chain | This Chapter, Porter (1985), Liedtka et al. (2014) |
| | Value net | Bovet and Martha (2000), Parolini (1999) |
| | SWOT analysis | Armstrong (1982), Barney (1995) |
| | Capabilities analysis | Harris and Lenox (2013) |
| Mini-tour focused on financials | DuPont analysis | This Chapter |

## 9.3.1   Understanding Customers

When attempting to understand customers, there exist two aspects to consider – the activities that lead to a purchase decision and the activities resulting from the purchase made. The former is best covered using the *customer decision journey* framework of Court et al. (2009) whereas the latter can be structured using the *journey map* framework (Liedtka et al. 2014).

### Framework—Customer Decision Journey

The customer decision journey framework identifies three stages that each potential customer, the so-called lead, goes through. First, once the lead has identified a need to be satisfied, a pain to be alleviated, or a gain to be sought after, they enter the *consideration set* stage. They are actively contemplating making a purchase. To move to the second stage, the *purchase decision* stage, the lead actively evaluates the offerings part of the consideration set. To do so, they draw from a diverse set of information, for example, brand trust, advertisements, social media comments, or embracement by influencers, to name just a few. A purchase decision is made. Finally, the offering is delivered and the lead, which has become at this stage a customer, is entering the *offering delivery experience* stage by using the purchase. This either results in a happy customer, potentially returning to the purchase decision stage or, if unhappy, returning to the initial consideration set stage.

**Example** Consider a student interested in buying a tablet or laptop for use in classrooms. Based on exploratory observing and ethnographic interviews, the persona Peter is looking for a device that helps him get three key jobs done, that is,

– search the internet for classroom course-related topical information,
– take notes during classes and exchange them with fellow students, and
– write and submit assignments.

Table 9.2 illustrates how insights gained from observing multiple students of the persona type Peter buying a tablet or laptop are mapped onto the customer decision journey and transformed into knowledge.

## Framework—Journey Map

The journey map framework represents a sequence of steps the customer performs after having purchased a product or service. For each step, the following information is depicted:

(1)  Customer journey step name
(2)  Description of the activities performed
(3)  Identification of the responsible party
(4)  Pre-conditions that must be met
(5)  Rational outcome from the activities performed
(6)  Emotional state after having performed the activities
(7)  Follow-up step or steps, depending on the outcome

When used during the learning step, the journey map focuses on the actual customer journey, whereas if it is used during the designing step (as described in Chap. 10), the focus is on the ideal customer journey.

Validating journey maps is best done by gallery walks. Gallery walks describe the customer journey using a walk-by gallery of posters illustrating the individual steps of the customer journey. Stakeholders are invited to visit the gallery and give constructive feedback. The feedback received is then used to revise and update the journey map.

**Table 9.2**  Illustration of how the customer decision journey framework can be used to structure insights and derive knowledge

Observation type and focus area

Observations identify that Peter starts defining his consideration set, that is, a list of possible laptop and tablet computers, for getting his jobs done by defining must have and nice to have decision criteria. These criteria can be grouped into two categories, for tablets and for laptops respectively, in decreasing order of their relevance to Peter.

| Tablet decision criteria | Laptop decision criteria |
| --- | --- |
| Brand | Battery lifetime |
| Features, like add-on keyboard, or pencil | Size and weight |
| Storage capacity | Connectivity (WLAN, Bluetooth, 5G, etc.) |
| Price (value for money) | Storage capacity |
| Usability | Price (value for money) |

Purchase decision

From all offerings that made it into the consideration set, Peter considers the following four sources of information to make his final purchase decision:
(1) Feedback from friends, and fellow students, especially which devices they rely upon
(2) Positive and negative user feedback found through Google on the internet
(3) Ease of use of the ordering process, availability, and delivery speed
(4) Evaluation of the prioritized features identified during building the consideration set

Offering delivery and experience

Peter forms his opinion about how satisfied he is with the purchase made considering the following criteria:
(1) On-time delivery without any hassles
(2) Usability of the selected product as expected and defined by his decision criteria, focusing on perceived usability, rather than hard facts
(3) Satisfaction with the results obtained from using the purchased tablet or laptop, with respect to the three jobs-to-be-done, primarily focusing on time saving and perceived quality of results

## 9.3.2  Identifying Capabilities and Resources

To gain knowledge related to the capabilities and resource of a firm, the value chain framework, introduced by Porter (1985), still is best practice. Proponents of the strategy design school, as described in Chap. 1, advocate using the SWOT analysis framework, as it is much simpler. But the SWOT analysis framework adds a judgmental component to knowledge, which is something that should be refrained from, in the context the learning stage in the design thinking strategy process.

### Framework—Value Chain

The value chain framework describes the individual activities and their sequence that a firm performs to create value for its customers and stakeholders. It focuses on the internals of the firm. The value chain of each firm

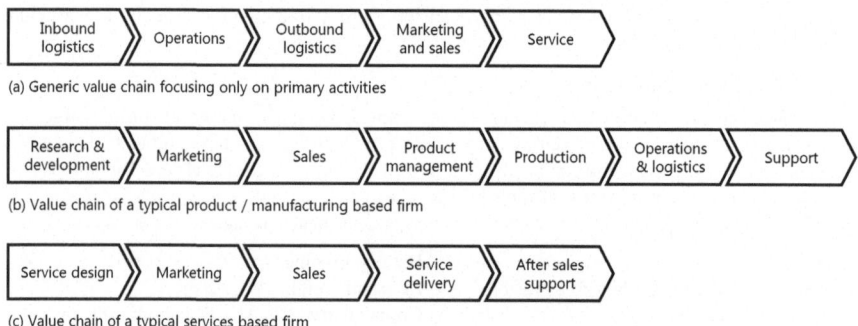

(a) Generic value chain focusing only on primary activities

(b) Value chain of a typical product / manufacturing based firm

(c) Value chain of a typical services based firm

**Fig. 9.1**  Typical sequence of activities defining the value chain of a firm

should, according to Porter (1985), be distinct. The value chain framework documents for each activity

- what the *activity* is delivering, both in terms of result and in terms of value,
- who the *key players* involved are, and
- what core *capabilities* (skills, resources) are required to perform the activity.

Figure 9.1 illustrates the generic structure of the value chain, (a) as proposed by Porter, (b) for a typical manufacturing company, and (c) for a typical service company.

### 9.3.3  Comprehending Financials

The challenges faced when trying to understand financial insights and derive knowledge is avoiding falling into the numbers trap. Learning is related to gaining knowledge that goes beyond simple numbers, as, for example, expressed by key performance indicators. The DuPont tree framework supports qualitative knowledge generation from insights by decomposing a firm's profit into its components. Knowledge can be modeled from insights for each of these components.

**Framework—Dupont Tree**

The DuPont tree framework decomposes the profits of a firm into its components. There exist different variations of the framework. The decomposition

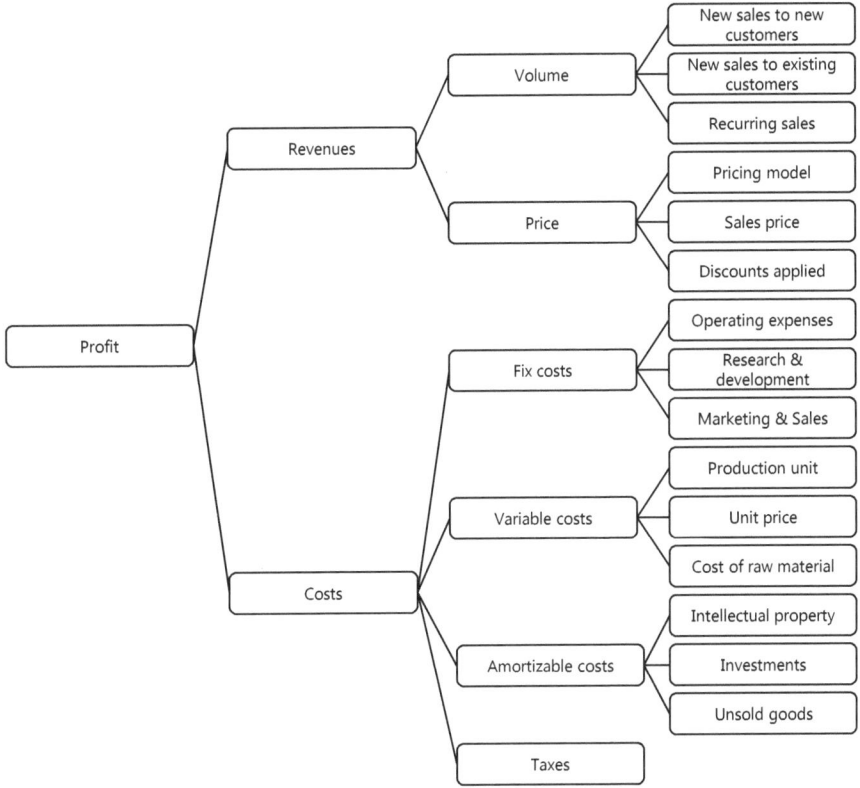

**Fig. 9.2** DuPont tree framework variation for use in the context of the business model design layer of the strategy design process

shown in Fig. 9.2 works best in the context of understanding different cash-flows of a firm in the strategy context and turning them into knowledge that can be exploited in the designing step of the strategy design process.

## 9.4  Mapping and Clustering Insights to Gain Knowledge

Knowledge is gained by building a model of the observed insights, or insights learned from interviews, focus groups, or secondary research, through mapping and clustering them onto the chosen frameworks by following four steps:

(1)  *Selecting a reference point* in the chosen framework, for example, the customer jobs-to-be-done element of the detailed business model framework.

(2) *Identifying all insights* that relate to the selected reference point, for example, all jobs-to-be-done that have been observed and that customers have expressed in ethnographic interviews and focus groups.

(3) *Deriving knowledge* from the insights by applying one or more of the following five activities:

    (i)    *Associating observations* with elements of the chosen framework, for example, by relating the brand to the customer relationship element in the detailed business model framework.

    (ii)    *Rephrasing insights using shared language*, ensuring that parties with diverse backgrounds gain a mutual understanding of the knowledge derived, for example, using the notion of news feeds to describe blogs, magazine articles, Twitter threads, etc.

    (iii)    *Calibrating knowledge* to ensure proper levels of breath and depth of knowledge by relating diverse insights to each other and simplifying them, for example, defining customer segments around the categories baby boomers, generation X, and millennials.

    (iv)    *Identifying interdependencies* between observations, such as relationships between sales prices and the costs of the goods sold, when selling interchangeable goods.

    (v)    *Formulating abstractions* by generalizing the insights gained, such as transforming insights related to the access to blogs, social media posts, and chat bots, into knowledge related to how customers use and value digital communication channels.

(4) *Identifying and documenting correlations and causalities* between the generated knowledge, for example, by relating identified jobs-to-be-done to appropriate customer segments.

Learning focuses on transforming insights into knowledge relevant to the designing step D of the strategy design process. The intentions behind the observations made are extracted and consolidated. This allows giving them context and putting them into perspective. Successful learning is about finding a compromise between depth and breadth of knowledge. Frameworks support finding that compromise.

**Example** Figure 9.3 illustrates how the mapping and clustering process transforms insights (figure a), left) into knowledge (figure b), right) using the customer and value proposition elements of the detailed business model framework from observing the grocery stores industry.

It is important to focus on the retained strategic focus, when learning from mini-tours and in-depth observations derived insights. Depending on the strategic focus, knowledge can be classified into any of the four categories:

| Customer Segments | Customer Relationship |
|---|---|
| • Parents with children<br>• Single people doing their daily shopping based on their short-tern preferences<br>• Individuals shopping based on prepared shopping list<br>• Elderly people visiting as part of their socializing activities<br>• Individuals having forgotten something | • Visiting a nearby grocery store<br>• Small talk with the cashier<br>• Asking for and getting help from store staff to find a given item<br>• Buying promotional items<br>• Using coupons to get rebates<br>• Avoiding long queues by using automated check-out |
| Customer Jobs-To-Be-Done<br>• Feeding a family<br>• Satisfying desires for specific dishes at home<br>• Buying specific items that households are out of<br>• Saving money by taking advantage of special offers<br>• Saving time when doing grocery shopping | |

a) observations / insights

L.2 to L.4

| Customer Segments | Customer Relationship |
|---|---|
| • Shoppers having a list what they need to buy<br>• Shoppers who walk through the isles and buy on impulse<br>• Shoppers who visit the grocery store as part of their routine rather than because of specific needs | • Grocery store no further than 10 min. walk or drive from home<br>• Friendly and helping staff<br>• Customer stickiness through promotions and advertised rebates<br>• Choice of check-out (cashier, automated check-out) |
| CustomerJobs-To-Be-Done<br>• Finding items that are on the list<br>• Seeing items that arouse a special desire<br>• Saving time by being as efficient as possible<br>• Selecting items to buy focusing on saving money | |

b) knowledge

**Fig. 9.3** Illustration of the learning process transforming insights related to customer jobs-to-be-done (a), into knowledge (b), focusing on the grocery stores industry

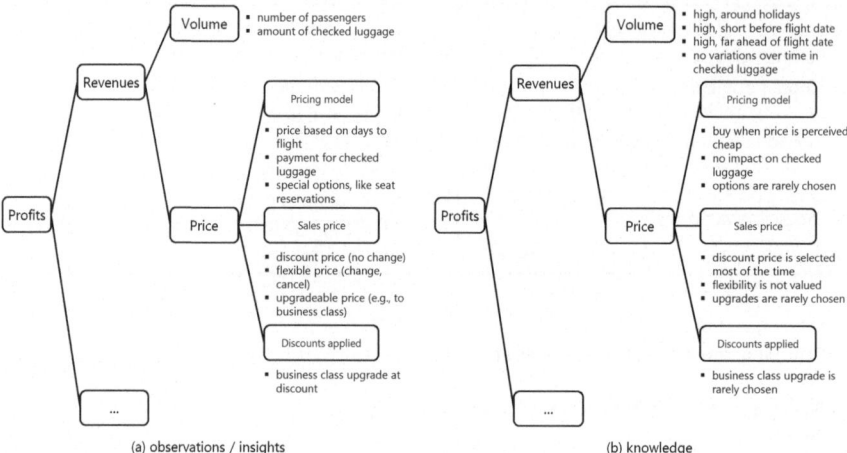

**Fig. 9.4** Deriving knowledge (**b**) from cash-flow based insights (**a**) using the DuPont tree framework for a discount airline

(1)  Identifying customer jobs-to-be-done.
(2)  Understanding the use of existing products and services, as well as support services, such as after-sales support.
(3)  Apprehending core capabilities, skills, and resources used.
(4)  Recognizing cash-flows, both on the revenues and cost side, as well as their occurrence over time.

Aligned with the focus taken during the observing step, learning focuses primarily, if not exclusively, on the chosen strategic focus dimension. Knowledge along the other dimension is only collected if and when needed during the designing step. This allows avoiding non-value adding analysis.

Consider, for example, a firm that wants to compete on satisfying unmet customer jobs-to-be-done. Observing and learning its current capabilities, may hinder identifying unmet needs that cannot be addressed with existing capabilities.

**Example** Figure 9.4 illustrates how knowledge can be gained from analyzing cash-flows of a discount airline under a financial strategic focus based mini-tour.

**Example** Think about a traditional watch manufacturer with expertise in designing and assembling automatic watches[1]. The observing step may have identified the customer need for battery driven watches, called quartz watches[2], especially for watches that customers only wear during special occasions and they do not want to reset the time and calendar each time anew. Primarily focusing on the capabilities would never allow identifying that new customer segment, as their needs cannot be met with existing automatic watch manufacturing capabilities.

---

[1]An automatic watch is a mechanical watch in which the natural motion of the wearer provides energy to run the watch, making manual winding unnecessary.
[2]A quartz watch is a watch that uses a battery-driven electronic oscillator that is regulated by a quartz crystal to run the watch.

**Table 9.3** Sample set of assumptions underlying the outcome of the observe process steps L.2–L.4 illustrated in Fig. 9.3

| Assumption category | Framework element | Formulated assumption | Validation approach |
|---|---|---|---|
| Observation | Customer segment | Customers in a given customer segment buy only what is on their shopping list | Ask customers after having paid how many items they have bought not on their shopping list |
| Observation | Customer segment | There exists a customer segment primarily buying on impulse | Ask customers at the entry what they want to buy and check after paying if they actually bought what they said |
| Modeling | Customer relationship | Grocery store must be no farther away than 10 minutes of walk or drive | Survey customers for their home address |
| Correlation/ Causality | Customer jobs-to-be-done | Customers are flexible on how to check-out, if it helps them save time | Provide different check-out options and check how often the faster ones are preferred over the slower ones |

## 9.5 Formulating and Validating Assumptions

The knowledge gained from the learning step is only as good as its validity. Therefore, it is important at this early stage of the business model layer to validate what has been learned so far.

The first step in ensuring soundness of the gained knowledge is formulating assumptions underlying it. Assumptions can be classified into three categories:

(1) Assumptions behind observations, for example, that the informants interviewed were representative of the personas they detail.
(2) Assumptions made during the modeling of the insights, for example with respect to abstractions made.
(3) Assumptions underlying the correlations and especially causalities related to the gained knowledge.

> **Example** Table 9.3 illustrates assumptions underlying the observed insights and learned knowledge from the grocery example in Fig. 9.3. Assumptions are classified into the three stated categories, as well as associated with the corresponding elements of the used framework.

Once assumptions have been formulated (step L.3), validations are defined and performed (step L.4). In contrast with statistical hypothesis testing, assumption validation is not related to getting a result with a great $t$-statistic. In most cases, such an approach would be infeasible due to the qualitative nature of the assumptions.

Even if statistical tests were possible, they would be of little value, especially when putting them into perspective of the costs and efforts required to perform them. The primary goal of testing is identifying what must hold for the assumptions to be true. This may be achieved through asking additional questions, conducting secondary research, or relying on classical surveys. The aim should be on understanding what could invalidate the assumptions made.

In some cases, it may not be possible to validate specific assumptions formulated with the input from the observing step with reasonable effort. In this case, giving feedback to what questions to address in upcoming observing steps, is most appropriate. This is especially the best approach early on during the observing step, after passive observations (step O.2) or ethnographic interviews (step O.3).

If an assumption can neither be validated, nor invalidated, and the underlying knowledge may be relevant to the designing step, it should be retained with an additional flag indicating its unverified status. This may be the preferred approach if the knowledge is hard to validate and its relevance during the designing step yet to be defined.

Experience helps chose the most appropriate validation approach based on compromising between effort and relevance.

## 9.6   Timeline and Required Skills

The timeline of the learning step is tightly linked to the one of the observing one. Experience has shown that spending 2/3 of the time on observing and 1/3 on learning provides sound results.

Best results are obtained when the individuals doing the observations, interviews, focus groups, and secondary research, are also in charge of the associated learning steps, up to and including assumption formulation. Ideally, the outcome of step L.2, that is, modeling insights leading to knowledge, should be cross-checked by, and discussed with, all strategy team members. The reason for this approach is twofold. First, it ensures a common understanding of the knowledge gained amongst all strategy team members. Second, it assures that personal biases are identified and addressed early in the process.

All assumptions identified during the learning step need to be discussed and prioritized by the strategy team, before starting their validations. Best results are obtained when the decision takers are performing the validations, or at least are actively involved in testing. It is that participatory role of decision takers that ensures that they believe in the outcome of the strategy design process. It also helps onboarding decision takers early on in the thinking arguments and reasonings behind the new or revised strategy.

# References

Armstrong, J. S. (1982). The value of formal planning for strategic decisions. *Strategic Management Journal, 3*(3), 197–211.

Barney, J. B. (1995). Looking inside for competitive advantage. *Academy of Management Executive, 9*(4), 49–61.

Bovet, D., & Martha, J. (2000). *Value nets: Breaking the supply chain to unlock hidden profits.* Hoboken, NJ: Wiley.

Christensen, C. M., Hall, T., Dillon, K., & Duncan, D. S. (2016a). Know your customers' "jobs to be done". *Harvard Business Review, 94*(9), 54–62.

Christensen, C. M., Hall, T., Dillon, K., & Duncan, D. S. (2016b). *Competing against luck: The story of innovation and customer choice.* New York, NY: HarperCollins Publishers.

Court, D., Elzinga, D., Mulder, S., & Vetvik, O. J. (2009). The consumer decision journey. *McKinsey Quarterly* (3).

Diderich, C. (2017). Initiating the strategy process using design thinking. *Change Management Strategy eJournal, 9*(8). SSRN: https://ssrn.com/abstract=2927941 or http://dx.doi.org/10.2139/ssrn.2927941.

Harris, J. D., & Lenox, M. J. (2013). *The strategist's toolkit.* Charlottesville, VA: Darden Business Publishing.

Liedtka, J., Ogilvie, T., & Brozenske, R. (2014). *The design for growth field book.* New York, NY: Columbia University Press.

Mootee, I. (2013). *Design thinking for strategic innovation.* Hoboken, NJ: Wiley.

Osterwalder, A., & Pigneur, Y. (2010). *Business model generation.* Hoboken, NJ: Wiley.

Osterwalder, A., Pigneur, Y., Bernarda, G., & Smith, A. (2014). *Value propositon design.* Hoboken, NJ: Wiley.

Parolini, C. (1999). *The value net: A tool for competitive strategy.* Hoboken, NJ: Wiley.

Porter, M. E. (1979). How competitive forces shape strategy. *Harvard Business Review, 57*(2), 137–145.

Porter, M. E. (1985). *Competitive advantage.* New York, NY: The Free Press.

Spradley, J. P. (1979). *The ethnographic interview.* Long Grove, IL: Waveland Press.

van Assen, M., van den Berg, G., & Pietersma, P. (2009). *Key management models* (2nd ed.) Harlow, UK: Pearson Education.

Verganti, R. (2009). *Design-driven innovation.* Boston, MA: Harvard Business Press.

# Shaping the Strategy by Designing Business Model Prototypes

# 10

> *Design isn't just about making things beautiful; it's also about making things work beautifully*—Prof. Roger Martin

The first step towards designing the firm's future strategy has been made by defining its strategic focus, that is, selecting the high-level direction along which the firm wants to compete and differentiate. The designing step of the business model layer writes the play to perform on the strategic stage. Designing is about generating novel ideas and combining existing knowledge in a novel way to describe how the firm will conduct its business and compete in the future. Designing is about creating options for the future around the firm's strategic focus. Designing is also about transforming those options into detailed business model prototypes that can be validated. The designing step is where the crucial creativity happens during the strategy design process. It is the most challenging step. Many ideas initially look encouraging, but most are a challenge to transform into prototypes. Just because an idea or a prototype of an idea looks promising to its designers, does not mean it will be accepted by customers or prospects.

Successful ideation is more about the quality of ideas than it is about the quantity, especially when aiming at disrupting. Only mediocre strategists focus primarily on quantity. Ensuring qualitatively superior ideas requires

- *creative people* with diverse backgrounds and interests that are open-minded and embrace the challenge of questioning the status-quo, and
- *strong collaboration* between creative people with a common goal of designing the next great strategy.

Strategy designers must not perceive time as an enemy. Creativity is not time constrained. Creativity is not about speed. There does not exist a correlation, let alone a causality, between time spent on identifying novel ideas and their quality. This is valid both ways, spending tool little or too much time.

Designing ideas and prototypes of the firm's future detailed business model is about the firm and not about competitors. Although strategy is inherently a relative

© Springer Nature Switzerland AG 2020

C. Diderich, *Design Thinking for Strategy*, Management for Professionals, https://doi.org/10.1007/978-3-030-25875-7_10

game in a competitive landscape, the designing step takes an absolute and firm-focused approach. A novel idea on how a firm can be desirable, deliver feasible products or services, and secure financial viability, itself lays the foundation for a competitive advantage in the target industry. Focusing on competition during ideation results in mediocre, incremental strategies. It also leads to benchmark thinking rather than differentiation. Potential adjustments to the business model for competition occur during the competition layer of the strategy design process (see Chap. 12).

## 10.1    Designing Objectives

The goal of the designing step of the strategy design process is to develop multiple testable prototypes of the firm's target detailed business model. Designing focuses on what is new and/or what is different relative to the firm's current detailed business model. Designing is exploratory and relies on divergent thinking. It is based on the two activities *ideating* and *prototyping*.

$$business\ model = ideas + prototypes$$

In strategy design, ideation and prototyping are so intertwined that it makes little sense to consider them as separate process steps, as is the case in design thinking applied to generic problem solving or product development. Prototyping in strategy is a mental activity, while in other areas of design thinking application, prototyping relates to physically building prototypes or mock-ups. Ideation is mostly about combining existing ideas to create something new or innovative.

**Example** Consider the idea of using the cash register of a 24/7-attended gas station as a human operated ATM available round the clock. Neither the cash register, nor the concept of an ATM, is new. However, the resulting idea of using the cash register as a human operated ATM is innovative. It gives bank customers access to cash on their account without the bank having to install additional expensive ATM hardware or branches open 24/7. In addition, by being human operated, it increases the trust factor and reduces the fear of customers being robbed. Finally, it provides value to the gas station owner because it reduces the amount of cash in the register, thus diminishing the cost of transporting cash to the bank and the risk of being robbed.

Designing starts by looking for novel ideas or new combinations of existing knowledge that business can transform into prototypes which deliver value to customers and the firm. The equation

$$innovation = new\ idea\ or\ new\ combination\ of\ existing\ knowledge$$
$$+\ value\ resulting\ from\ willingness\ to\ pay$$

applies.

## 10.2  The Designing Process

The designing process D centers around four key activities:

(1) *Describing the existing detailed business model* as starting point for innovation (if the goal is to develop a disruptive strategy or a strategy for a start-up, this first step can be left out).
(2) *Generating multiple novel ideas or new combinations of existing and/or novel ideas*, targeting the strategic focus elements of the detailed business model.
(3) *Building prototypes* of both the strategic focus and the offerings (OVP and OPS) elements of the detailed business model.
(4) *Completing the remainder of the detailed business model*, driven by desirability, and addressing feasibility, and viability.

Similar to the interaction between observing (O) and learning (L), validating should follow each designing activity, that is, executing the validating process V, described in Chap. 11. It is best practice to build and validate distinct prototypes for testing the desirability, feasibility, and viability of the ideas at hand. Depending on the outcome of validation, the prototypes, or even the underlying ideas, may need adjustment, amendment, or even be discarded. To avoid non-value adding activities, strategy designers should prioritize ideas, prototyping of strategic focus-based elements, and full prototype designs, by focusing on

– the *expected contribution to success* of the strategy, in decreasing order, and
– the *complexity of validation*, starting with concepts that are easy to validate.

Depending on the strategic focus, validating desirability (customer and offering strategic focus), feasibility (capabilities strategic focus), or viability (financials strategic focus) should be prioritized.

---

**Process D—Designing**

D.1  Documenting the current detailed business model (optional, when aiming for a disruptive strategy or in the start-up context)

D.2  Iteratively selecting a target population identified during the observing (O) and learning (L) steps on which to focus the design (target populations should be prioritized in decreasing order of their expected relevance to the strategy to be designed)

D.3  Based on the knowledge created during the learning process (L), combined with the outcome of the environmental analysis (E), generating novel ideas and/or combining existing knowledge in novel ways, by concentrating on the strategic focus elements of the target detailed business model

V    *Validating the ideas generated by confronting them to the real world*

D.4   Designing prototypes related to the strategic focus and offerings
        elements (OVP and OPS) of the detailed business model
V       *Validating the designed prototypes*
D.5   Completing the prototypes by designing the remaining elements of the
        detailed business model
V       *Validating the completed detailed business model prototypes*
D.6   Aggregating the designed prototypes from multiple target populations,
        if sound
V       *Validating the aggregated detailed business model prototypes*

## 10.3  Documenting the Current Detailed Business Model

Unless the goal is to develop a disruptive strategy or a strategy for a start-up, that is,
a firm that does not yet exist, the first step of the designing process is describing the
firm's current detailed business model. This activity can be subdivided into two
parts, that is,

– *describing each element* of the firm's detailed business model, being as neutral
  as possible, and
– *documenting the relationships and causalities* between the different elements of
  the firm's detailed business model.

**Example** Figure 10.1 illustrates a possible outcome from the first step (D.1) of the
designing process for a domestic news agency, such as the Australian Associated Press,
Deutsche Presse-Agentur, Kyodo News, Schweizer Depeschenagentur, or The Canadian
Press.

If a firm implements its current strategy through multiple separate business units,
a detailed business model should be used for describing each distinct business unit.
Each business units may be considered a separate firm with its own business model
and strategy.

Describing the detailed business model of a firm is teamwork, best performed in
a classical workshop setting using Post-it® notes or Stattys on a pin-wall with a
detailed business model poster attached to it. Strategy designers may use filament,
colored needles, notes with distinct colors, or color marks to document
relationships.

One challenge faced during the detailed business model documentation step is
identifying the appropriate degree of detail. Although there is no single right answer
to this question, less is usually more. Experience has show that an A0-sized poster

| Customer Segments (CS) | Customer Relationship (CR) | Value Proposition (OVP) | Competitive Advantage Activities (KAC) | Cost Advantage Activities (KAC) | Outsourced Activities (KAO) |
|---|---|---|---|---|---|
| • Daily print newspapers<br>• Online news sites<br>• Search engines<br>• Communication departments of firms, associations, and governments | • Dedicated sales representative<br>• 24/7 news desk reachable via various channels (phone, e-mail, chat, etc.) | • Objectively written news<br>• Fast delivery<br>• Global coverage through partner network<br>• Fee related to circulation, rather than news gathering effort | • Exclusive partnerships with foreign news agencies | • Broad domestic coverage | • Foreign news coverage<br>• System integration into customer's news system |

| Customer Jobs-to-Be-Done (CJ) | Customer Delivery(CD) | Products & Services (OPS) | Perishable Resources (KRP) | | Capital Resources (KRC) |
|---|---|---|---|---|---|
| • Inform readers about latest newsworthy events<br>• Be the first to report the news<br>• Ensure global news coverage<br>• Save costs by avoiding own presence at news events<br>• Direct traffic to online platform | • Automatic feed into customer's news system<br>• Access through protected web site | • Written news ready to be published | • News | | • News handling and distribution IT system |
| | | | **Labor (KRL)** | | **Skills (KRS)** |
| | | | • Journalists with local knowledge for domestic coverage<br>• News editors | | • Local knowledge<br>• Coverage scheduling and prioritizing |

| Revenues (FR) | Cost Structure (FC) |
|---|---|
| • Subscription fee based on circulation/access | • Journalists, news editors, and sales representatives salaries<br>• News handling and distribution IT system development and maintenance<br>• Fee for distributing partner news agencies' content |

**Fig. 10.1**   Detailed business model of a domestic news agency

of the detailed business model, combined with typical $4 \times 3$ inch (or $10 \times 7$ cm) notes, is sufficient to document most firms' detailed business models.

Seasoned line managers and strategy designers should jointly be able to document the detailed business model of the firm or an independent business unit in no more than one to two days of collaborative work.

## 10.4   Generating Innovative Ideas

Generating innovative ideas is at the core of the creativity phase of the strategy design process. Successful ideation is difficult, and successful innovation even more. Innovations are ideas that customers are interested in and willing to pay for.

### 10.4.1   Selecting a Target Population

Ideation starts by selecting a target population. A target population, in the traditional sense of design thinking, is a persona related to a customer segment, and one or more jobs-to-be-done. For example, a coffee shop may select the persona Jenney, a young mother wanting to socialize with acquaintances, as a target population to consider. In design thinking for strategy, the target population may alternatively be a technology, such as, blockchain or artificial neural networks, or a specific capability, such as a cost-efficient implementation of passive mutual funds or supply chain management. Strategy designers identify target populations to consider

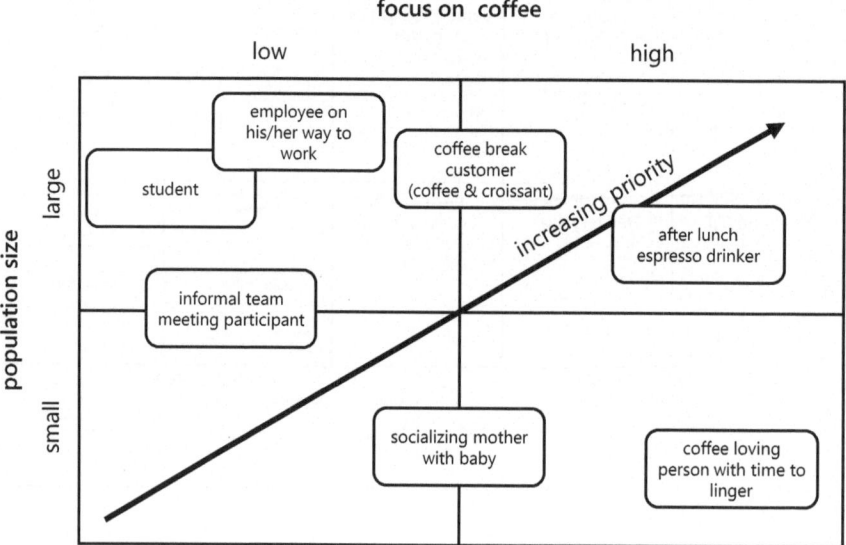

**Fig. 10.2** Characterization and prioritization of possible target populations related to a customer focused coffee shop strategy

during the observing (O) and learning (L) processes. If new target populations are identified during the designing step, the processes O and L should be performed on them before moving to using them in designing. Target populations should be prioritized before being considered. Prioritizing is a key skill that any strategy professional must exhibit.

> **Example** Figure 10.2 illustrates a possible prioritization for a typical customer-centric coffee shop strategy, focusing on relevance and coverage.

### 10.4.2  Ideation

After having selected a target population, ideation focuses on knowledge related to the target population, gained during the learning step (L) (Chap. 9) and the insights identified by the environmental analysis process (E) (Chap. 6), to come up with novel ideas or novel combinations of knowledge along the firm's strategic focus. The current detailed business model, documented during step D.1, serves as the basis to define ideas in terms of change from the status-quo, rather than in a greenfield way. Innovations are based either on novel ideas or new combinations of existing ideas and knowledge. More often than not do innovations come from re-combining existing knowledge, rather than from something completely different and new.

Experience shows that the first idea is usually not the best one and is often even a quite poor one. Initial ideas can be improved upon by applying one or more of the following transformation techniques, that is,

- *magnifying*, thinking bigger and/or smaller,
- *multiplying*, extracting value from scaling,
- *inverting*, trying out the opposite of the original idea,
- *stretching*, extending one specific property of the original the idea,
- *compacting*, reducing the impact of one specific property of the original idea, and
- *decomposing* the idea into its components and deriving sub-ideas of value from one or more components.

The focus of ideation differs depending on the firm's chosen strategic focus.

*Customers strategic focus* Ideation focuses on identified jobs-to-be-done and designing novel value propositions and associated offerings that address them. The emphasis is on existing or potential needs to alleviate felt pains or generate sought after gains.

*Offerings strategic focus* Ideation focuses on what is traditionally called inventions. It aims at creating new needs or addressing existing jobs-to-be-done in a novel and superior way. Inventions may be novel technologies, new user experiences, new models, or new processes, that create needs that do not yet exist.

*Capabilities strategic focus* Ideation focuses on how firms can use existing capabilities, such as, skills, resources, processes, or capital, to create novel offerings and provide novel value to customers.

*Financials strategic focus* Ideation based on a financials strategic focus means looking at willingness-to-pay, cash-flows and their timing, as well as costs.

Not all generated ideas make it to the idea prototyping step D.4. However, this does not mean that they should be discarded right away. A list of unused, but not fully discarded, ideas should be kept on the sideline for future reference. At the end of the ideation step D.3, ideas retained for step D.4 are validated by confronting them with the real world.

### 10.4.3 Typical Examples of Ideas

Rather than looking backward and presenting case studies on how firms successfully innovated in the past or failed to do so, this section presents a sample set of building block ideas, partially formulated as questions, which can be used for bootstrapping design. Ideas are approached in a generic, rather than target industry or target population specific, way. It is the strategy designers' role to identify those ideas that matter in their specific context and use the transformation approaches described, to generate new ideas and transform them into innovations. The individual ideas are classified along the four possible strategic focuses of a firm.

### 10.4.3.1  Customers and Their Jobs-to-Be-Done

Ideation around customers and their jobs-to-be-done is related to better understanding what attributes of a strategy they consider of value to them. Four main areas of ideation can be identified.

*Properties* What are the properties that customers focus on when aiming at addressing their jobs-to-be-done? Are they related to price, that is, getting a fair solution at a cheap price? Or do customers value time and wish to get their jobs-to-be-done done as quickly as possible? How do customers value the need for flexibility, getting similar jobs done using the same offering? Or, do customers value the ease of use, for example, by not having to read a user's guide? Different answers to these questions will lead to different customer segments that firms can best address with diverse strategies and associated value propositions.

*Decision takers* More often than not, the end users and the decision takers or check writers are not the same person. They have related, but different, jobs-to-be-done on their minds. This is especially the case in a business-to-business environment. The decision taker, writing the check, may aim at competitively priced solutions satisfying a pre-agreed upon catalogue of functionalities, whereas the end user focuses on product quality and unique features supporting their specific jobs-to-be-done. Innovation, in this case, is related to identifying a compromise that maximizes the combined utilities of the different parties involved.

*Communication channels* Customer-centric ideation should consider opportunities around communication channels used to interact with customers before, during, and after sales. Typical buzzwords in this context are multi-channel and omni-channel. The creativity challenge is identifying and designing the right channels for the targeted customers and avoiding those channels that customers to do not value. More is not always better. A key mistake to avoid is assuming that a digital channel satisfies all customers at all time and is the only medium of interaction needed.

*Delivery* An offering is only as good as its delivery. When offering physical goods, firms may consider different delivery mechanisms, like home delivery, pick-up at a specific location convenient for the customer, or leveraging third-party locations for delivery. Not all products need to be sold in stores. Stores may be only used for showcasing. When the offering is intangible, like a service, firms could design legal structures or technology-based delivery channels around specific jobs-to-be-done. Not every service delivery needs a physical presence. For example, a maintenance service may be delivered using video and audio to advise the end-user on how to solve a given problem.

### 10.4.3.2 Products- and Services-Focused Ideation

In many strategies, even those not primarily focusing on offerings, products and services characteristics play an important role.

*Usage* Typically, an offering may be used once, like a frozen pizza, or multiple times, like a drill. One-off use offerings may lead to recurring purchases. They may also lead to cheaper production, increasing competitiveness, especially, if customers have a given job-to-be-done to address only once.

*Types of jobs-to-be-done* Are customers looking for an offering that addresses a specific jobs-to-be-done, that is potentially unique, or are they considering a generic jobs-to-be-done such that a single offering may address multiple similar jobs-to-be-done?

*Choice* When looking for a solution to their jobs-to-be-done, customers may value options. Are the customers looking for an offering that solves a specific job-to-be-done in the best possible way or do they prefer a more generic offering that can address multiple similar jobs-to-be-done? Perhaps the customers prefer to adapt the offering for a specific job-to-be-done themselves, for example, a drill that can be re-configured as a screwdriver? The customers may prefer buying a portfolio of tools at a discount. Alternatively, the customers may be interested in an offering configured for their specific needs and preferences, like computers bought configured to the customer's specific wishes. Do the customers want to be offered the best solution for a job-to-be-done or prefer a choice between multiple reasonable solutions? How much do the customers want to be involved in choosing versus outsourcing the choice to the vendor?

*Support* Firms may take different approaches to after-sales support. It could be part of the offering or be sold separately as a one-time service or on an annual subscription basis. Alternatively, firms can offer case-based support and charge a fix price or on a time basis. Support coverage may be included in the ideation around and after-sales service. Return policies are another area of support that can result in innovative strategies, such as subscription-based business models in which customers can exchange a product each time their need changes.

*Substitutes* Customers have a preference for how to get their jobs done based on their experience. For example, a customer wanting to buy a home, their job-to-be-done, but lacking sufficient funds, may look for a mortgage. Really innovative offerings-based strategies are designed around disruptive alternatives to solve a given job-to-be-done. A home leasing offering could result in an innovative strategy in a competitive homeowner market.

Many more options to innovate around offerings exist. When ideating about new products and service, it is important not to forget the overall strategic context. Ideation must relate to all aspects of a firm's strategy, from being desirable, through ensuring feasibility and viability, to exhibiting a positioning advantage allowing to prosper in a given industry environment.

### 10.4.3.3  Leveraging Capabilities

The third area for innovation involves the firm's capabilities. Rather than inventing new capabilities, ideation is mostly framed around recombining existing capabilities, in an innovative way.

*Skills*  Firms can strategically leverage their capabilities in four primary areas, that is,

– *technologies*, such as computer science technologies or technologies around production, like synthesizing substances used in the pharma industry,
– *business processes*, like procurement, supply-chain management, or after-sales support,
– *knowledge*, like patents, intellectual property, or in-depth subject matter expertise, and
– *access to customers*, in which the firm relies on its unique capabilities to connect with customers, such as through eco-systems or platforms.

*Usage of skills*  When ideating about how to re-configure existing skill to design a new or modernize an existing strategy, firms can use any of the following four approaches:

(1)  Identify *new target populations* for which the firm can use its existing capabilities to address customer jobs-to-be-done.
(2)  Identify *new jobs-to-be-done in existing target populations* that the firm can successfully address by using existing capabilities in a novel way.
(3)  Develop *new features to add to existing offerings* to better leverage existing capabilities.
(4)  Exploit *opportunities for selling more existing offerings to existing customers* by leveraging existing capabilities to reduce costs and/or increase rational and emotional value for customers.

Although more specific, capabilities-based ideation should also consider how the firm can leverage its ability to satisfy regulatory and legal constraints to offer additional customer value. Customers may perceive the firm satisfying regulatory constraints by 120%, rather than 100%, as providing differentiating value.

### 10.4.3.4  Ideation Around Financials

Recent years saw many innovations around pricing and pricing models. Ideation in pricing is related to aligning revenues with customers' perception of them. Some firms have built their strategy solely around innovative—which needs not mean cheap—pricing. There exist at least five key areas in which pricing ideation can lead to value for both the firm and the customer.

*Frequency of payments*  Should payments be one-off or recurring? When aiming for recurring payments, are they down-payments or do they offer a true subscription value? What are the advantages, from a customer perspective, in the offered models?

*Timing of payments*  The timing of payments is important, for the firm and from a customer perspective. Timing has a significant impact on the firm's working capital requirements. From a customer perspective, timing may impact the perceived trust in the firm and its offerings. Typically, payments may be made at the time of purchase, after use, or on a pay-as-you-go basis.

*Units of value*  An often poorly understood concept in pricing innovation is the concept of unit of value. Different customer segments may prefer different units of value for the same offering. Typical units of value are lump sums, quantity- or volume-based, time-based, cost-based, usage-based, performance-based, or as degressive units.

*Payer*  As noted previously, the end user is not always the paying customer. Firms can exploit this distinction by designing pricing models that explicitly differentiate in the payment model between different stakeholders. The most common such model is an advertisement-based model. In another model, airport shops pay the passenger landing taxes, either with or without requiring a purchase, because the airport generate valuable leads for the shops.

In addition to classical efficiency-based ideation, innovating around costs can consider the surrounding environment to generate value and differentiation by

- exploiting available purchasing power with suppliers,
- leveraging economies of scale by perfecting and/or centralizing purchasing processes, or
- joining forces or outsourcing procurement to obtain better conditions.

The ideas presented are only crumbs, of which many more exist, especially those targeted to specific industries. Successful strategy designers excel at

- identifying novel ideas based on observing and learning,
- combining ideas to create innovations,
- putting innovations into a business model context and a subsequent strategy, and
- understanding that only a small fraction of ideas will make it into a successful strategy.

## 10.4.4   Ideation Tools

There exist many tools for steering the ideation process. The most popular one is brainstorming (Osborn 1963). It helps generate many ideas to evaluate and prioritize. A key characteristic of brainstorming is avoiding commenting on and criticizing ideas during the generation process, which proponents identify as a high-value property.

---

### Tool—Brainstorming

Brainstorming is an old ideation tool. It was first proposed in 1942 by Osborn (1963), who argued that one of the main barriers to creative productivity was that most ideation sessions failed because their primary focus was on evaluation. He described the problem as *driving with the brakes on*. Brainstorming aims at addressing this flaw by focusing in a first step solely on producing lists of ideas which can be subsequently evaluated and further processed. Osborn defined brainstorming as an ideation method based on four guiding principles:

(1) *Criticism is ruled out.* Adverse judgment of ideas is to be withheld until after the brainstorming session.
(2) *Freewheeling is welcome.* Brainstorming encourages diverse approaches to generate innovative ideas, allowing wild and unusual ideas.
(3) *Quantity is preferred.* Brainstorming is based on the premise that the more ideas are formulated, the greater the likelihood of useful ideas being among them.
(4) *Combinations of ideas are appreciated.* Brainstorming explicitly encourages combining previously formulated ideas to turn them into better ideas. Brainstorming sessions often incorrectly prohibit this last guiding principles because it may be perceived as criticism, although constructive criticism.

Although deferred judgment is a central element of brainstorming, Osborn made clear that judging ideas is important, but mixing ideation and judgment is not the best way to move forward. Over the years, researches revised, adapted, and sometimes diluted the brainstorming method (Timpe 1987; Furnham and Yazdanpanahi 1995; Dugosh and Baulus 2005; Kohn and Smith 2011; Gregersen 2018). It nevertheless remains the most used ideation technique.

---

Generating a large number of ideas often creates an adverse selection bias. Therefore, brainstorming is useful for incremental innovation, but does not address the challenges faced by radical innovation (Verganti 2009).

A less well-known alternative to brainstorming better suited for radical innovation is anti-conventional thinking (ACT), developed by Baumgartner (2015). It focuses on depth rather than breath during ideation.

---

**Tool—Anti-conventional Thinking**

As a fervent opponent of brainstorming, Baumgartner (2015) introduced the anti-conventional thinking (ACT) method, based on three critical flaws found in brainstorming sessions:

(1) Brainstorming focuses on quantity rather than quality. Consequently, brainstorming results in long lists of mediocre and similar ideas.
(2) Brainstorming prohibits criticism. Criticism is a key tool to disrupt common sense. Thus, the no-judgment rule in brainstorming, leads to most ideas being conventional.
(3) Brainstorming is a highly structured approach, leading to a tunneled view, often missing disruptive ideas.

ACT is a method for generating highly creative ideas, focusing on depth rather than breath of thinking. It is modeled after the way creative people, such as artists, writers, or composers, think and collaborate. It is also based on scientific research around how the human brain operates. ACT is a six-step approach:

(1) *Make a situation transcendental.* Rather than start ideation with a common-sense question, ACT starts with describing the initial question in an unconventional way.
(2) *Play with the situation.* Before starting with ideation, ACT scrutinizes and rephrases the question in distinct ways. The goal is to gain multiple perspectives of the challenge at hand, from both a rational and an emotional perspective.
(3) *Formulate an extreme goal.* Rather than focus on a challenge to solve or a question to address, ACT takes a constructive stance and requires formulating an extreme goal, that may or may not be achievable, to address the challenge or question at hand.
(4) *Build a creative vision.* Only in the fourth step are ideas formulated with the aim to achieve the stated extreme goal. All generated ideas are tested through mind games. Too conventional and non-viable ideas are rejected on the spot. Ideas are criticized constructively with a focus on the boring parts of ideas. Criticism is formulated as questions that encourage discussion and debate. The creative vision is build through trial and error around iterative ideation and mind-validations.
(5) *Build an action plan.* Although not formally part of ideation, ACT requires that participants formulate an action plan describing what to do with the designed creative vision.

(6) *Do it*. ACT explicitly includes a step requiring participants to take action based on the ideated vision and design plan.

ACT is preferred as ideation tool when the goal is to generate disruptive rather than incremental ideas. It is also a valid alternative to brainstorming if participants do not believe in the assumption that quantity will lead to quality, or that critiquing inhibits creative thinking.

Another tool for conducting innovation sessions is the LEGO® SERIOUS PLAY® method (Kristiansen and Rasmussen 2014; Smith and Meyerson 2015; Blair and Billo 2016; Smith et al. 2017). It allows ideating around 3-d models, rather than relying on 1-d voice and text or 2-d drawings. It helps make a 3-d print of the mind in a very efficient way.

## Tool—LEGO® SERIOUS PLAY® Method

The LEGO® SERIOUS PLAY® method is a facilitated thinking, communication, and problem-solving tool for use with organizations, teams, and individuals. It draws on extensive research in business strategy, organizational development, psychology, and learning. It allows participants to explore and deal with genuine issues and challenges in real time by relying on metaphors, figures of speech, and narratives. Extensive sharing of meanings helps everyone feel ownership of the ideas expressed.

The LEGO® SERIOUS PLAY® method is a four-step process that supports constructive ideation:

(1) *Defining challenge*. The facilitator formulates the challenge to address.
(2) *Building models*. Participants build LEGO® models representing their reflections on the challenge, that is, new ideas and new combinations of existing ideas.
(3) *Sharing meaning*. Participants share the meaning and story behind the models their built.
(4) *Reflecting insights*. The team reflects on the insights gained from the individual models and stories to derive and prioritizes a list of kept ideas.

Participants use LEGO® bricks, Duplos, and mini-figures to create visual models that express their thoughts, reflections, and ideas, to address the posed challenge. Storytelling helps participants share meanings and gain insights.

There exist many more ideation tools, some generic, some specific to typical situation or design team structures (Eppler et al. 2014). They all require teamwork. Multiple half-day sessions, ideally away from the day-to-day environment,

**Table 10.1** Sample output from a creativity session focusing on the three largest populations of a coffee shop

| Target population | Prioritized ideas |
|---|---|
| Employees on their way to work | (1) Allow pre-ordering through an app<br>(2) Offer coffee on a subscription basis<br>(3) Introduce a loyalty program based on regularity of consumptions rather than on quantity |
| Coffee break customer (coffee and croissant) | (1) Offer discounts based on occupancy<br>(2) Offer mini-meeting rooms for a fee<br>(3) Offer coffee and croissant as a bundle<br>(4) Offer coffee flavors based on weather |
| Students | (1) Offer small working spaces including power outlets and internet access<br>(2) Offer discounts for multiple coffees per session<br>(3) Offer discount coupons for use during specific hours<br>(4) Offer cheap lunch packages<br>(5) Design a "bring a friend" offering that includes coffee for two and pastries |

especially phones, laptops, and e-mail access, should be conducted to support ideation. Deadlines and time pressure are adversaries to creativity. In addition, it is not possible to command and/or deliver creativity 24/7. Multiple shorter ideation sessions provide far better results than fewer longer sessions do. Strong unbiased moderation is key to supporting the process, allowing for sufficient creativity and idea exchanges while avoiding any derailment.

> **Example** Table 10.1 illustrates a sample output from a creativity session focusing on coffee shops.

## 10.5  Transforming Ideas into Business Model Prototypes

Transforming ideas into workable business model prototypes proceeds in three iterative steps separated by testing and validation activities.

(1) The idea is transformed into a description of one or more elements of the detailed business model that relate to the strategic focus and value proposition.
(2) The remainder of the detailed business model elements are designed and causalities between elements defined.
(3) Prototypes stemming from different target populations and/or ideas are aggregated and commonalities, especially in the non-strategic focus related elements, identified.

The goal of any prototype is to test the validity of the underlying business model. It is important to design distinct prototypes, depending on whether to validate the

– *desirability*, focusing on customers and their jobs-to-be-done relative to the value propositions and offerings,
– *feasibility*, aiming at validating the required capabilities to deliver upon the promises made in the value proposition, or
– *viability*, targeting financial aspects of the business model ensuring sustained profitability in excess of the costs of capital.

The first step focuses on describing the details of the strategic focus elements, the value proposition (OVP), and products and services (OPS) elements related to the ideas considered.

**Example** Think of a retail bank that identifies mobile phone addicted young people that just entered the workforce, as a possible target population. Ideation may identify that this target population wants to be able to track their finances whenever and wherever. The prototyped offering elements could be a mobile app showing the customer's account balances well as a decomposition of recent payments into typical budget categories that allow reviewing spending with minimal hassle. Figure 10.3 shows a typical customers strategic focus-based business model prototype for the described retail bank.

A first consistency check is ensuring that any value proposition characteristic is provided by at least one offerings characteristic. Second, the value proposition characteristics must be related to the strategic focus characteristics. In the case of a customers strategic focus, this means, showing desirability, for a capabilities strategic focus, ensuring feasibility, and for a financials strategic focus, confirming viability. In the case of an offerings strategic focus, all three characteristics, that is, desirability, feasibility, and viability, are part of the first validation step in business model prototyping.

Once the strategic focus-based elements of the designed business model prototype have been validated, the second step of prototyping aims at completing the two, respectively, three remaining components of elements of the detailed business

| Customer Segments (CS) | Customer Relationship (CR) | Value Proposition (OVP) |
|---|---|---|
| • Young people (around 18 – 30 years)<br>• Mobile phone addicted<br>• Recently entered workforce, i.e., having limited savings and focusing primarily on consumption | • Young and trendy brand<br>• Word-of-mouth referrals for new customers, supported by social media advertisements<br>• Communication via mobile phone only | • Customer sees account balances 24/7 on mobile phone app<br>• App automatically clusters booked expenses into typical categories<br>• History based budgeting tool supports controlling spending<br>• Backup communication channel providing human problem solving, supporting trust building |
| **Customer Jobs-to-Be-Done (CJ)** | **Customer Delivery(CD)** | **Products & Services (OPS)** |
| • Want to have full control over their spending<br>• Want to be able to check 24/7 what their account balance is<br>• Want to understand where they spend their money | • Mobile phone app (iPhone and Android)<br>• Backup communication via chat and VoIP using artificial intelligence technology supported by human, in case of problems | Mobile phone app offering:<br>• Real-time account balance<br>• Expense clustering<br>• Budgeting tool based on past spending, including soft and hard limits<br>• Link to a free credit card<br>• Integrated invoice payment services |

**Fig. 10.3** Customers strategic focus-based business model prototype derived from the idea of servicing mobile phone addicted banking clients

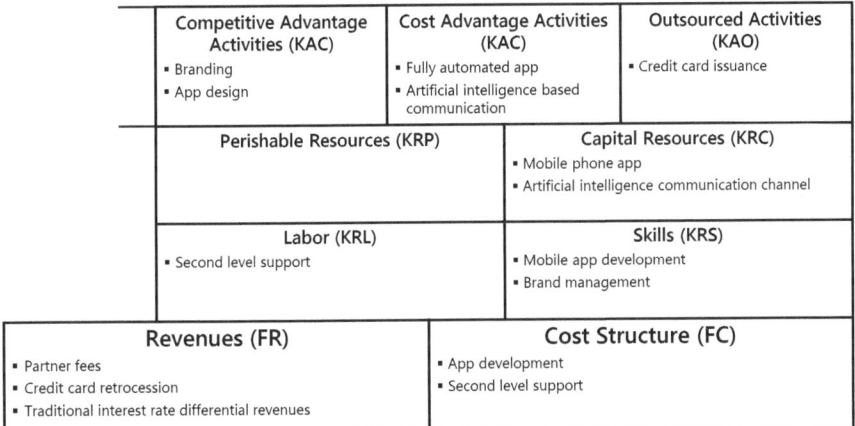

| Competitive Advantage Activities (KAC) | Cost Advantage Activities (KAC) | Outsourced Activities (KAO) |
|---|---|---|
| ▪ Branding<br>▪ App design | ▪ Fully automated app<br>▪ Artificial intelligence based communication | ▪ Credit card issuance |

| Perishable Resources (KRP) | Capital Resources (KRC) |
|---|---|
| | ▪ Mobile phone app<br>▪ Artificial intelligence communication channel |

| Labor (KRL) | Skills (KRS) |
|---|---|
| ▪ Second level support | ▪ Mobile app development<br>▪ Brand management |

| Revenues (FR) | Cost Structure (FC) |
|---|---|
| ▪ Partner fees<br>▪ Credit card retrocession<br>▪ Traditional interest rate differential revenues | ▪ App development<br>▪ Second level support |

**Fig. 10.4** Capabilities and financial elements of the detailed business model prototype for a retail bank focusing on mobile phone addicted young clients

model. As with the strategic focus-based elements of the business model prototype, the aim is not only on describing the elements of the detailed business model, but also on describing their interdependencies.

**Example** Figure 10.4 illustrates the remaining elements of the detailed business model of the retail banking example in Fig. 10.3.

Transforming ideas into prototypes is a team activity requiring the participation of strategy designers knowledgeable in the different steps of the value chain. A workshop setting with all team members in one room is recommended. A neutral, usually external, moderator is preferred to avoid conflicts of interest and ensure that everyone gets a say. During prototyping, multiple teams may work in parallel on multiple prototypes. Table 10.2 illustrates a typical timetable for a one-day prototyping workshop focusing on two specific ideas addressing the same target population. Typical prototyping workshops for firms with multiple target audiences last between three and five days. Workshops covering process step D.4, focusing on designing the strategic focus related elements of prototypes, should be separated from workshops focusing on the remaining elements of detailed business model prototypes. The time in-between covering the two topics (process step D.4 and process steps D.5/D.6) should be used for validation. During steps D.5 and D.6, subject matter experts may be invited to the prototyping workshops.

**Table 10.2** Sample timetable for a single target population customers strategic focus based prototyping workshop

| Activity | Duration | Lead/participants |
| --- | --- | --- |
| (1) Presenting the ideas serving as the basis for prototyping | 15 min | Moderator |
| (2) Prototyping the customer elements CS, CJ, CR, and CD, and offerings elements OVP and OPS of the target detailed business model | 1 h | Teams of 3–4 strategy designers |
| (3) Presenting of the prototypes designed by the teams and critiquing by the other participants | 10 min each | Each team, plenum |
| (4) Reviewing and updating of the prototypes, based on the feedback received | 30 min | Same teams |
| (5) Presenting of the revised prototypes and performing a second round of critiques | 10 min each | Each team, plenum |
| (6) Classification of insights related to the different elements of the prototypes into the categories: agreed, to validate, rejected | 30 min per prototype | Moderated plenum session |
| (7) Third round of prototype updating based on the classified insights gained and their mutual understanding | 1 h | Different teams |
| (8) Developing validation experiments for the insights that need validation and assigning the validations to specific team members | 1 h | Same team as in third round of prototyping |
| (9) Agreeing upon the next workshop date, location, participants, and deliveries | 10 min | Moderator |

## 10.6   Aggregating Prototypes Stemming from Multiple Ideas

The last step in prototyping detailed business models (D.6) is about aggregating multiple prototypes and identifying commonalities and complementarities. Participants need to ensure consistency throughout the aggregated prototypes. Depending on the design outcomes, multiple competing prototypes may be derived. In this case, the decision on which detailed business model to base the strategy may be taken at a later stage, for example, after the validation step or even after reviewing the prototypes in their competitive environment, as described in Chap. 12. In some cases, it may even be sound to keep multiple detailed business model prototypes and associate them with distinct business units or even distinct legal entities within the same group. A typical multi-strategy situation may occur if one detailed business model prototype focuses on a low-cost based value proposition and another one on a premium value proposition, focusing on complementary customer segments.

Best practice shows that complementary detailed business models should never be merged because this would result in a diluted positioning in the market and diminish the value propositions.

## References

Baumgartner, J. (2015). *Anticonventional thinking: The creative alternative to brainstorming*. Erps-Kwerps, Belgium: JPB.

Blair, S., & Billo, M. (2016). *How to facilitate meetings & workshops using the LEGO serious play method*. London, UK: ProMeet.

Dugosh, K. L., & Baulus, P. B. (2005). Cognitive and social comparison processes in brainstorming. *Journal of Experimental Social Psychology, 41*(3), 131–320.

Eppler, M. J., Hoffmann, F., & Pfister, R. A. (2014). *Creability*. Stuttgart, Germany: Schäffer-Poeschel Verlag.

Furnham, A., & Yazdanpanahi, T. (1995). Personality differences and group versus individual brainstorming. *Personality and Individual Differences, 19*(1), 73–80.

Gregersen, H. (2018). Better brainstorming. *Harvard Business Review, 96*, 64–71.

Kohn, N., & Smith, S. M. (2011). Collaboration fixation: Effects of others' ideas on brainstorming. *Applied Cognitive Psychology, 25*(3), 359–371.

Kristiansen, P., & Rasmussen, R. (2014). *Building a better business using the lego serious play method*. Chicester, UK: Wiley.

Osborn, A. F. (1963). *Applied imagination: Principles and procedures of creative problem-solving*. New York, NY: Scribner.

Smith, J. L., & Meyerson, D. (2015). *Strategic play: The creative facilitator's guide*. Tunbridge Wells, UK: Wordzworth Publishing.

Smith, J. L., Meyerson, D., & Walling, S. (2017). *Strategic play: The creative facilitator's guide #2: What the duck!*. Tunbridge Wells, UK: Wordzworth Publishing.

Timpe, A. D. (1987). *Creativity*. New York, NY: Facts on File Publications.

Verganti, R. (2009). *Design-driven innovation*. Boston, MA: Harvard Business Press.

# Managing Uncertainty Through Experiment-Based Validation

**11**

> *Our success at Amazon is a function of how many experiments*
> *we do per year, per month, per week, per day*—Jeff Bezos

Two types of mistakes can often be observed in strategy design processes. The first is, executives believing that they know their customers better than customers do know themselves. This leads to offerings being developed that nobody wants, or nobody is willing to pay for. Typical examples were the DAT[1] offering music in a digital format on an analog cassette medium, or the personal digital assistant Newton,[2] Sony's Betamax video tape format, Nissan's Murano CrossCabriolet, McDonalds McWings, New Coke, Microsoft Zune, or 3D television sets, just to name a few. CB Insights (2018) identified that 42% start-ups fail because there is no market need for their products and services. The second big mistake often observed, on the opposite end of the scale, is decision takers only being willing to decide if they are 100% convinced that change will be successful. They ask for validation after validation in an attempt to remove any business risk from their decision making. This can often be traced back to significant above average risk aversion, a trait that is at odds with successfully leading organizations into the future.

The challenge any strategy designer is facing is finding a compromise between not moving ahead fast enough and taking too much risk. The validating step of the business model layer aims at supporting decision takers by providing enough evidence to convince them that it would be very hard to come up with additional evidence that would make them change their mind. The focus of the validation step is on reducing risk down to an acceptable level. That does not and should not mean completely eliminating risk. Validating assumptions is forward-looking and should not be confused with hypothesis testing, as known from statistics theory, which, by nature, is related to extrapolating the past into the future.

---

[1]DAT = Digital Audio Tape, developed by Sony and introduced in 1997, but never embraced by the music industry.
[2]Newton was introduced by Apple in 1993. It failed to attract enough customers due to its high price and problems with its handwriting recognition feature. It was retracted from the market in early 1998, after Jobs returned to Apple.

© Springer Nature Switzerland AG 2020

C. Diderich, *Design Thinking for Strategy*, Management for Professionals,
https://doi.org/10.1007/978-3-030-25875-7_11

## 11.1  Validating Objectives

While designing the detailed business model using process D (Chap. 10), choices are made based on sound assumptions. Although strategy designers believe in the assumptions they make, that does not necessarily mean that these are true. Assumptions should be validated earlier rather than later during the strategy design process. Validating assumptions early avoids possible costly mistakes later on.

There exist two possible approaches to validate formulated assumptions:

(1) The assumption is translated into a *quantifiable hypothesis* that can be tested using statistical methods and algorithm (Kuehl 2000). This is the typical approach used in academic research. Statistical hypothesis testing relies on historical data and is inherently backward-looking, making it inappropriate for achieving the forward-looking goal of strategy design.

(2) The assumption is *related to a design decision* in the detailed business model, either directly to a specific element, to a relationship between elements, or to the environment. Rather than relying on historical data to validate the assumption, judgmental insights are gathered up to the point where the marginal added knowledge from any additional insight on the validity of the assumption becomes nearly zero. *Judgmental validations* take a forward-looking stance and aim at getting first-hand insights. Judgmental insights go beyond a simple yes or no answer. Getting to an 80% certain positive answer with sound explanations is preferred over a 90% certain answer without such explanations.

In design thinking, judgmental validation comes to application. Validation is used as a decision support tool rather than a truth finding mechanism. The focus is on managing the uncertainty related to strategy decisions rather than getting them exactly right. Strategy includes, by definition, a certain degree of uncertainty. Statistical hypothesis testing can be used in a simplified form, if the strategic goal is to extrapolate the past into the future. This may be the case for incremental or fast-follower strategies.

## 11.2  The Validating Process

The assumption validating process V is by its nature a forward-looking confirmatory process. Even though it is not trivial and requires significant experience to formulate sound testing experiments, validation is systematic and straightforward. This sometimes leads to the fatal mistake being made, believing that validation can be performed by junior staff or outsourced to external consultants. As the primary goal of validation is to support strategic decision making, decision takers involved in the decision-making process should also be involved in the validation process. If executives have heard first hand from a customer that a given idea, the assumption, is valid or invalid, they will be much more confident into the associated decision than if that information would have been relayed to them by a third party.

To ensure success of validations, it is necessary to educate and coach the decision takers to be involved in validating assumptions, especially, if they did not participate in the designing step of the strategy design process. Having decision takers perform mock-up experiments jointly with strategy designers provides the necessary confidence for both parties that the validation outcome can be relied upon. A final caveat to note is that validation activities are not and must not be considered sales activities. The goal is getting objective feedback that helps deciding and not convincing someone that a given idea or a new offering is great. This is hard but must be ensured at all cost. Therefore, owners of an idea should not play a leading role in its validation.

### Process V—Validating

V.1   Formulating assumptions

V.2   Classifying assumptions based on their impact on desirability, viability, and feasibility and prioritizing them relative to their relationship to the strategic focus, their design impact, their validation costs, and their strategy risk

V.3   Designing experiments to validate/invalidate the assumptions made

V.4   Performing the designed experiments

V.5   Deriving consequences from the experiments' outcome on the designed detailed business model, its elements, and connections

V.6   Testing the desirability, viability, and feasibility of the detailed business model as a whole using a top-down perspective

Although described as a separate process, validation is an integral part of designing the detailed business model. Each time an assumption, whose validity is key for the next design decision, is made, it should be validated as soon as sound. Validations should not solely be scheduled at the end of the business model layer. Sometimes it is sound to prefer an early validation of an idea using a simplified experiment over a full-fledged experiment at a later stage. This is especially the case when the validity of the assumption significantly impacts subsequent design decisions. The full-fledged validation of such assumptions may be combined with related validations later on. Determining the timing and effort required for each validation activity is a key skill a strategy designer must exhibit. It is a trade-off between

– the *impact* of the assumption on subsequent design decisions in the detailed business model,

– the *cost*, with respect to time and money, of performing an experiment to validate the assumption, and

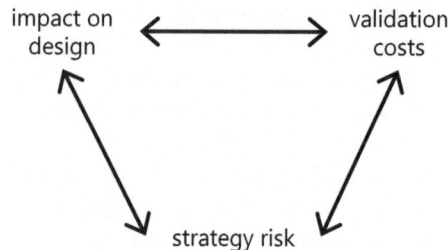

**Fig. 11.1**  Trade-offs relevant for prioritizing assumption validations

- the permitted *strategy risk* or uncertainty underlying the resulting detailed business model,

as illustrated in Fig. 11.1.

If the design impact is minor and/or validation costs are excessive, it may sometimes be sound to accept the associated strategy risk and not validate a given assumption, only validate it later in the strategy design process, or combine its validation with one or more related assumptions at a later stage.

## 11.3   Formulating Assumptions

The first step in reducing the uncertainty behind the designed detailed business model or elements of it, is to formulate assumptions. An assumption is a belief related to the future that may or may not be true. There exist three kinds of assumptions to consider:

(1) *Element-based assumptions.* Are the descriptions of specific elements of the detailed business model valid?
(2) *Relationship-based assumptions.* Are the descriptions of the relationships between elements of the detailed business model valid?
(3) *Externality-based assumptions.* Are the assumed causalities between externalities and the descriptions of the specific elements of the business model valid?

Each detailed business model is based on many assumptions. To avoid unnecessary analysis, only those assumptions

- that have a *material impact on the validity* of the detailed business model, and
- for which the *confidence is insufficient to accept the consequences* of an incorrect decision

| Customer Segments (CS) | | Value Proposition (OVP) |
|---|---|---|
| • Young adults entering the workforce<br>• Home owning families<br>• Tech savvy adults of all age | | • 24/7 access to funds via<br>  — wire transfer<br>  — gas station<br>• 100% online mortgage handing process<br>• Overdraft facility based on extrapolated past cash-flows |
| Customer Jobs-to-Be-Done (CJ) | Customer Delivery (CD) | Products & Services (OPS) |
| • Wire transfers<br>• Month-end account overdrafts<br>• House financing relying on mortgages<br>• Cash retrieval and deposits at any time | • Mobile app<br>• Cash deposit retrieval at local gas stations | • Mobile app allowing<br>  — digital payments<br>  — on-line mortgage applications<br>  — accounts overdrafts<br>• Access to physical cash retrieval and deposit via local gas stations |

**Fig. 11.2** Excerpts from a prototype of a detailed business model describing a suburban retail bank developing a customer-centric purely digital strategy

should be validated. A typical detailed business model contains between 10 and 30 assumptions to be validated.

**Example** Consider a suburban retail bank having chosen an offerings-based strategic focus by becoming a pure digital bank. The strategic focus aims at making the live of customers easier through solely relying on technology, such as mobile apps, to deliver the offerings. Figure 11.2 illustrates excerpts of a prototyped detailed business model.

A typical element-based assumption to validate is "There are sufficient home owner families requiring mortgage financing in the suburbs covered by the bank (CS and CJ elements)." The assumption can even be extended to whether that customer segment, in addition to being large enough, is growing and currently under-serviced. A relationship-based assumption underlying the detailed business model is "The targeted customers (CS element) are willing to do all their payments (CJ element) via their mobile phone (OVP and OPS elements)." The assumption "Gas stations (externality) are willing to function as human serviced ATMs (OVP element) for a fee (FR element)" represents a typical externality-based assumption. The assumptions whether there is a market for mortgages, given a sufficiently large home owner customer segment, is a typical assumption that needs no validation, unless the neighborhood is very rich (externality) and its residents do not finance their home ownership through mortgages. Relationship-based assumptions are the most common ones to validate, followed by externality-based, and element-based ones.

## 11.4   Classifying and Prioritizing Assumptions

Not all assumptions are created equal. The relevance of an assumption depends on

- whether or not it *supports the firm's strategic focus*, and
- whether or not it is *relevant for* the firm's detailed business model to be *desirable*, *viable*, or *feasible*.

Once the assumptions have been formulated, they are classified in categories using the two-dimensional framework illustrated in Fig. 11.3, focusing on the primary impact on the success of the detailed business model with respect to desirability, viability, and feasibility, on the *x*-axis and on whether or not the assumption relates to the strategic focus on the *y*-axis. The classification helps distinguish between critical and non-critical assumptions. Non-critical assumptions that are found invalid can, in general, be fixed rater easily. Therefore, they do not require high priority attention, especially if they are hard or costly to validate. In addition, assumptions without any expected material impact on the detailed business model can often be ignored.

To ensure a cost efficient and effective validation, relevant assumptions classified in the same category should be prioritized based on two criteria, that is,

- what *effort*, in terms of time and money, is required to test the assumption, and
- how *significant* would the impact of a failed assumption test be on the validity of the detailed business model as a whole.

| | | primary impact on the success of the detailed business model | | | |
|---|---|---|---|---|---|
| | | desirable | viable | feasible | other |
| superiority impact | strategic focus related | high relevance<br><br>critical for success, if the strategic focus is on customers | high relevance<br><br>critical for success, if the strategic focus is on financials | high relevance<br><br>critical for success, if the strategic focus is on capabilities | should be reconsidered |
| | not strategic focus related | low relevance<br><br>can usually be fixed ex-post | low relevance<br><br>can usually be fixed ex-post | low relevance<br><br>can usually be fixed ex-post | can safely be ignored |

**Fig. 11.3** Classification of assumptions in categories based on their strategic relevance and impact on success

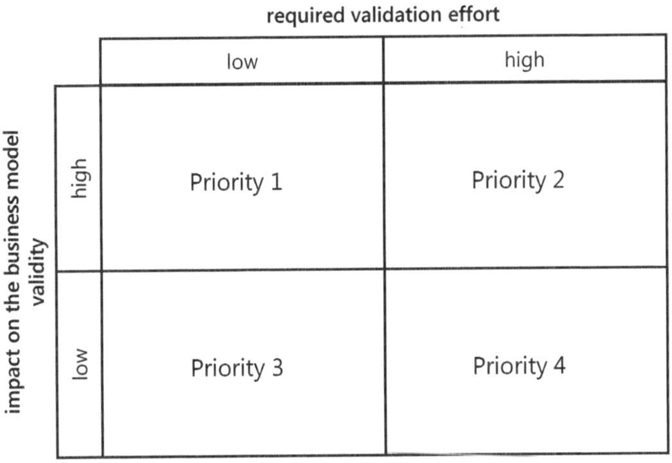

**Fig. 11.4** Matrix for prioritizing assumptions falling into the same category

Figure 11.4 illustrates a framework describing the resulting order in which assumptions belonging to the same category should be validated.

## 11.5   Designing and Conducting Experiments

Once assumptions have been formulated, classified, and prioritized, the creative work around validation starts. For each assumption or cluster of assumptions, an experiment must be designed. Similar assumptions may be clustered together and validated using a single experiment. Sometimes, multiple complementary experiments may be necessary to validate a single assumption.

An experiment to validate an assumption consist of five parts:

(1) A *closed-end formulation of the assumption* that allows for a yes or no answer, avoiding as much as possible a maybe answer.
(2) An *experiment to be performed* for finding out if the answer to the assumption is yes or no.
(3) A *representative and reasonably sized target informant population* on which to perform the experiment.
(4) A *measurement criterion* that translates the outcome of the experiment into a yes or no answer related to the assumption.
(5) A *threshold on the measurement criterion* that allows accepting or discarding the assumption.

**Table 11.1** Description of a sample experiment to validate the assumption that homebuyers are willing to contract their first-time mortgage via a mobile app

| Assumption category | − Customer strategic focus related<br>− Desirable |
|---|---|
| (1) Closed-end formulation of the assumption | Customers targeted are willing to contract their first-time mortgage via a mobile phone app without human interaction or support |
| (2) Experiment to be performed | Present informants a possible user-interface for contracting a first-time mortgage, focusing on the information they must provide on-line to process the application. Allow the informants to ask understanding questions around the process until they are sufficiently confident to have understood how the contracting process would work. Then ask the question if they would be willing to use such an app and follow the proposed process |
| (3) Target informant population | Home owners that have recently contracted a mortgage by visiting a bank branch. Initially select a population size of 25 and increase it by 10 additional informants until the experiment is conclusive or the target population size of 100 is reached |
| (4) Measurement criteria | Count as yes, all informants that answer the question with yes or maybe and count all others, including those unable to give a definitive answer, as no |
| (5) Decision threshold | − Accept the assumption if 80%, respectively 75% for population sizes larger than 25, of the informants questioned have been counted as yes<br>− Reject the assumption if 80%, respectively 75% for population sizes larger than 25, of the informants questioned have been counted as no<br>− Reject the assumption if an informant population size of 100 has not lead to a conclusive answer<br>− Add additional informants to the target population if the experiment has been inconclusive, according to the defined target informant population rule |
| Experiment characteristics | Cost: Low<br>Effort: Medium, due to the requirement to develop a possible user-interface prototype for a possible mortgage contracting application<br>Impact: High, as mortgages are perceived as a key offering of the digital bank |

**Example** Table 11.1 illustrates the description of a typical experiment used for validating assumptions regarding the digital retail banking example from Fig. 11.2. The measurement criteria can be defined in an incremental way, rather than as an absolute figure. Having at least 80% of positive or negative responses, is considered conclusive. In a first stage, 25 informants are questioned. If the outcome is not decisive, the threshold is slightly reduced, for example, to 75%, and the number of additional informants in the target population increased by 10, and so forth. Such an approach allows validating assumptions with minimal effort, as an increased effort is only needed when a heightened uncertainty exists.

Experiments should adhere to the 5 × 5 × 5 rule (Schrage 2014), that is, require no more than 5 weeks to be performed, cost no more than $5000 (or equivalent in local currency), and require no more than 5 strategy design team members and decision takers participation. For low effort and/or low impact assumptions, the rule may be simplified to 5 × 5 × 2, that is, no more than 5 days, $500, and 2 strategy designers and decision takers involved.

Designing experiments is a forward-looking and creative process. Academic insights into experiment development is often of lesser relevance due to its backward-looking nature. Using external support for designing, but not performing, experiments often proves to be of value as it allows for a fresh view and avoids potential confirmatory biases in the designed experiments. The primary goal of any experiment design should be on attempting hard to invalidate the to be tested assumption, rather than confirm its validity.

## 11.5.1  Typical Experiments

Although the space for designing experiments is limitless, typical experiments fall into one of the four categories, in decreasing order of their relevance, that is,

– feedback around mock-ups or prototypes,
– confirmatory interviews,
– split tests, or
– traditional surveys.

All experiments have in common that their outcome is only as good as their design. Enough time must be allocated for their development. Any experiment should be tested on a mock-up population before being administered to informants in the target population.

### 11.5.1.1  Mock-up or Prototype Feedback

Mock-up or prototype experiments present the informant a mock-up or a prototype of the assumption to be validated. Prototypes, whether physical or mental, are used instead of questions. The informant should be able to play around with the prototype and give feedback on its validity.

Mock-up based experiments are especially useful to validate offering features and distribution channels. They are regularly preferred over interviews to validate user experiences, as they avoid potential biases introduced by questioning.

**Example** Going back to the digital bank example illustrated in Fig. 11.2, testing if customers would be willing to buy their mortgage on-line, a key assumption behind the designed detailed business model, a mock-up-based experiment could be used. A sequence of screen masks would be presented to the informant to navigate through the mortgage application process to find out if such an approach would appeal.

### 11.5.1.2  Confirmatory Interviews

Confirmatory interviews, in contrast to explanatory or ethnographic interviews, focus on getting answers to close-end questions, rephrasing the assumptions to be validated. They focus as much on validating assumptions as on understanding the answers.

As with ethnographic interviews, confirmatory interviews start with putting the informant at ease to ensure that the answers are comprehensive and trustworthy.

Typical confirmatory interviews include questions along five dimensions:

(1) Dou you agree or disagree with the assumption?
(2) Why do you come to your conclusion? Which insights impact your decision most? Which insights did you discard or consider irrelevant?
(3) What would make you change your mind?
(4) What missing information could solidify your opinion?
(5) What attributes underlying the assumption were irrelevant to your decision making and could subsequently be ignored?

Answers to these questions allow not only testing the assumptions, but also understanding how the detailed business mode could be updated to better meet the formulated assumptions, if validated, or adjust it to address identified issues.

> **Example** Consider the assumption that customers want to be able to retrieve cash at any time from their bank account, as suggested in the example in Fig. 11.2. Assume that the informant does not agree with the assumption. He may comment his answer by indicating that what is important to him is the possibility to get cash early in the morning in order to be able to pay for a coffee on his way to work (his job-to-be-done) or get cash late in the evening to pay for the home-delivered pizza (his job-to-be-done). These insights may be used to rephrase the "at any time cash availability" assumption, by a 6am to midnight alternative or even add to the gas station cash withdrawal option, a pizza boy-based cash home delivery service. In addition, the closeness to the location to get cash may be described as more important than the nature of the location, that is, gas station. A grocery store with extended opening hours may be an acceptable alternative. Note that in contrast with the designing step, validating is not about coming op with alternative payment methods, but validating is about how and when customers want to retrieve cash.

### 11.5.1.3  Split Testing

Split testing experiments, either through simple A/B testing (Siroker and Koomen 2015) or more sophisticated multi-variate testing (Izenman 2008), are used when the assumptions lead to validating possible alternatives, rather than answering pure yes or no question. Split tests are commonly used to test assumptions around

- offering features,
- packaging and combination of characteristics, and
- pricing models.

> **Example** A split test may be used to find-out whether customers are willing to pay up-front, prefer payment in installments, or pay only after the product or services has been fully delivered.

Split test experiments are easy to design and allow going beyond a simple yes-no answer. They can be included in interviews or administered through surveys. On the downside, split-tests often lack by design the insights that can be gained from confirmatory interviews. Sometimes split tests may be complemented with confirmatory interviews, especially if initial results are inconclusive.

### 11.5.1.4  Surveys

Surveys are the most common and easiest to administer type of validation experiments. A large informant population can be reached with minimal effort. Even more important than in other experiment approaches, is the quality of the formulated question used to test the assumption. As the informant filling out the survey usually cannot be observed and ask understanding questions, the surveyed questions must be structured in a way that ensures honest and complete answering. Biases must be avoided. Confirmatory questions, rephrasing previous questions, should be included to test for consistency of the answers.

> **Example** Figure 11.5 illustrates a subset of a questionnaire administered via an online tool to validate the assumptions behind the detailed business model of the digital bank example in Fig. 11.2. Administering a validation survey for a digital bank strategy via an online platform introduces an informant selection bias that must be addressed, for example, by administering the survey to a random subset of informants through phone or via paper forms or by conducting interviews.

## 11.6  Validating Desirability, Viability, and Feasibility

Up to now, validation has focused in a bottom-up way, on individual assumptions behind elements of the detailed business model, relationships among them, and interdependencies with the environment. To ensure consistency of the designed detailed business model in a holistic way, the validation process V concludes with a set of top-down tests focusing on ensuring desirability, viability, and feasibility.

To validate desirability, viability, and feasibility, distinct experiments must be performed as these three test areas are complementary and only provide minimal overlaps. Significantly different approaches to experimenting are required. Validating that a give feature is desirable requires getting objective feedback from end-users, whereas validating viability of the same features is related to finding out if the purchase decision maker, rather than the end-user, is willing to pay for the

> *Please answer all survey questions honestly and focus on your preference, rather than your current behavior.*
> *Only select one answer. If multiple answers apply, select the most appropriate answer.*
>
> 1. In which category of potential customers would you classify yourself?
>
>    ❏ young adult, entering the workforce; ❏ homeowning family; ❏ family renting a home; ❏ tech savvy
>    adult; ❏ non-tech savvy adult; ❏ other
>
> 2. Which needs related to banking must a bank satisfy to enter your consideration set?
>
>    ❏ it must allow for submitting payments 24/7; ❏ it must support cash overdrafts at the end of the
>    month; ❏ it must offer competitive mortgages; ❏ it must offer access to cash outside of business hours;
>    ❏ it must offer access to cash 24/7; ❏ it must offer investment advice services;
>    ❏ it must be the cheapest in town
>
> 3. How do you prefer to interact with your bank and/or your banker?
>
>    ❏ via phone; ❏ by visiting a branch; ❏ if possible, through mail; ❏ via my PC at home/at the office;
>    ❏ via a mobile device I carry with me; ❏ via a trusted human, which may or may not be a bank employee;
>    ❏ none of the above options
>
> 4. ...

**Fig. 11.5** Excerpt from a survey used to validate assumptions around the digital bank detailed business model example illustrated in Fig. 11.2

feature. It is a mistake to assume that if a business model is desirable, it is automatically viable and feasible, and vice versa.

## 11.6.1  Validating Desirability

The desirability requirement of a detailed business model, that is, the offerings satisfy customer needs and support one or more of their jobs-to-be-done, can be validated by testing three high-level categories of assumptions:

(1) There exist enough customers in the targeted customer segments. The customer segments are expected to grow over time, or at least, not shrink.
(2) The firm can build a relationship with the targeted customers in a way that the firm's offering falls into the customer's consideration set.
(3) The value proposition offered by the firm covers enough attributes of the target customers' jobs-to-be-done to trigger a buying decision.

Validating these desirability assumptions can be done in a comparable way to testing other assumptions. Figure 11.6 illustrates the relationship between elements of a generic detailed business model and the assumptions. Competitive aspects of the desirability, for example, why a customer should favor the firm's offering over that of competitors, are addressed in Chap. 12.

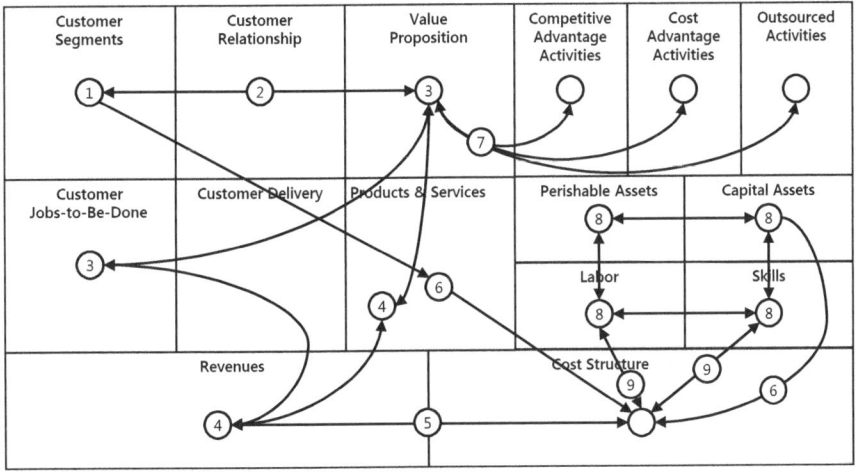

**Fig. 11.6** Nine typical assumptions underlying desirability, viability, and feasibility, of a typical detailed business model

## 11.6.2   Validating Viability

Not all desirable detailed business models are also viable. A business model is considered viable if customers are willing to pay for the offered value proposition a price that exceeds the costs of producing and delivering the offering, including costs of capital. Key assumptions to validate the viability are:

(4) Customers are willing to pay a given price for the offering to satisfy their jobs-to-be-done. This price allows the customer to perceive receiving sufficient value from the offered value proposition to trigger a buying decision.
(5) The expected revenues exceed the incurred costs, that is, the sales price is appropriate from the firm's perspective.
(6) A sufficiently large number of customers are willing to buy the offering and pay for it such that the investment made as well as fixed expenses are covered.

## 11.6.3   Validating Feasibility

To be successful, a firm must be able to deliver upon the promises made to their customers with the value proposition. It must be feasible for the firm to produce the offerings at the quality level expected by the customers. Unless the detailed business model is based on untested inventions, the firm aims at disrupting existing offerings, or the firm is completely inexperienced in the target industry, feasibility is often the least hard trait to ensure. The three most important assumptions to validate the firm's business model feasibility are:

(7) The firm is able to identify and perform the activities required to deliver upon the promises made by the value proposition, that is, address the jobs-to-be-done of the targeted customers.
(8) Sufficient resources are available at reasonable costs allowing the production and delivery of the offerings in the quality expected by the customers.
(9) The firm is able to use key assets and resources in an efficient and effective way, minimizing the risk of failing, to produce the offering in a desirable and viable way.

It is important to ensure that the assumptions behind desirability, viability, and feasibility remain valid not only at a given point in time, but throughout the lifetime of the detailed business model and associated strategy.

## 11.7  Risks to Avoid

At the end of the validation step, insights gained from experimenting should be fed back into the detailed business model layer of the strategy design process, mainly the design stage. Before doing so, three key traps must be avoided, that is,

- the false positive bias,
- the false negative bias, and
- the wrong data trap.

In statistics theory, the first two traps are called type I and type II errors respectively. Particular care must be taken when selecting an unbiased sample of informants to avoid all three traps in forward-looking business model validation experiments. One way of doing so, is having the experiment set-up and related data reviewed by an outside expert in experiment validation. This is especially important for those assumptions that make or break the detailed business model designed.

## References

CB Insights. (2018). Top 20 reasons why startups fail. *Research Brief.* https://www.cbinsights.com/research/startup-failure-reasons-top/.
Izenman, A. J. (2008). *Modern multivariate statistical techniques.* Heidelberg, Germany: Springer.
Kuehl, R. O. (2000). *Design of experiments: Statistical principles of research design and analysis.* Boston, MA: Duxbury-Thomson Learning.
Schrage, M. (2014). *The innovator's hypothesis.* Cambridge, MA: MIT Press.
Siroker, D., & Koomen, P. (2015). *A/B testing: The most powerful way to turn clicks into customers.* Hoboken, NJ: Wiley.

# Part V

# Exposing the Designed Strategy to the Competitive Environment

# Exploiting Findings from Game Theory to Succeed in a Competitive Environment

*If people do not believe that mathematics is simple, it is only because they do not realize how complicated life is*
—John von Neumann

Business is a high-stake game (Brandenburger and Nalebuff 1995). Strategy is about ensuring that the firm plays the right game in the right way. During the first two layers of the strategy design process, the foundation and the business model layers, the focus is on the firm. The third layer, the competition layer, aims at aligning the designed detailed business model with the competitive environment to finalize the strategy design. The firm's competitive advantages in the target industry are defined, either with respect to being different or being superior, where superior can mean cheaper. There exist multiple players, not directly under the control of the firm, that have an impact on success. One of them are competitors. Customers and their behaviors are another one. Key talents need also be considered, as they affect the competitive positioning. Strategy development requires to identify those players and exploit them to the firm's advantage or design counter-measures mitigating their potential negative impact. The firm's competitive advantage describes its unique positioning among all key players. In extension to traditional strategy schools embracing the competitive advantage approach, design thinking-based strategy development puts a strong focus on the role of the customer to competition.

## 12.1  What Competitive Advantage Means

Think about the last time you were buying a watch. What made you chose one brand over another? Or was your choice driven by features, style, size, availability? Or was your purchase an impulse decision? What was the job you wanted to get done with buying that new watch? Was it knowing the time, or was it more, or something different, like gaining status, tracking your fitness or having your e-mail around your

© Springer Nature Switzerland AG 2020                                                181
C. Diderich, *Design Thinking for Strategy*, Management for Professionals,
https://doi.org/10.1007/978-3-030-25875-7_12

wrist? These are all legitimate questions you have answered implicitly or explicitly when buying that new watch. Now put yourself in the shoes of a watch manufacturer, whether it is Apple, Blancpain, Rolex, Swatch, Tissot, or any other brand—their chief strategist, business developer, product manager, or even CEO. Wouldn't your job be much easier if you knew the answers to all those questions? The detailed business model describes how a firm operates and delivers value to its customers along its strategic focus. An absolute viewpoint, putting the firm at the center, is taken. The competitive advantage layer positions the firm, together with its detailed business model, in the competitive environment defined by its target industry and its players. A relative viewpoint is considered. Successfully competing requires understanding the different players' incentives and their threats and actions to achieve a competitive advantage themselves (Ghemawat 1997). Dynamic competitive analysis goes beyond the static analysis promoted by Porter (1980). It considers the evolution of the competitive advantage over time and looks at strategy as a game.

A successful strategy identifies and attains a competitive equilibrium among all involved players, putting the firm center stage. As such an equilibrium is transient in nature, strategy adjustments are needed over time. The competitive layer of the strategy design process defines the equilibrium, through making the competitive advantage of the firm explicit and pro-actively, rather than reactively, using game-theory, to anticipate potential changes in the competitive environment over time.

## 12.2   Understanding How to Compete

Even more than in the past, the success of any firm depends on its capabilities to differentiate itself from competitors in a way that customers perceive as superior and valuable. Traditional strategy scholars address the competitive positioning challenge from the firm's viewpoint. They take an inside-out view to answering the question "what makes the firm superior to its competitors". Superiority can be achieved through competing on differentiation, competition on price, or positioning in a niche segment (Porter 1980). The key challenge with this approach is that it assumes a seller driven market and relegates the customers' view on value to the second row.

More recently, novel approaches focusing on customers and their jobs-to-be-done have been developed (Christensen et al. 2016). They put the customers and their needs, their felt pains, and sought-after gains center stage. The competitive positioning is derived by mapping the firm's value proposition underlying its offerings to those needs. This approach works well in an environment with limited competition, for example, resulting from disruptive characteristics of the offering.

Although inherently sound, both approaches to competitive positioning fail to answer two key questions in an explicit and holistic way:

(1)   Why should a customer prefer the firm's offering over that of its competitors?
(2)   How will competitors react to the firm's competitive positioning over time?

Answering these questions leads to identify two primary approaches to succeed in a competitive environment, that is, either *being different* from competitors or *being superior* to competitors. The value proposition describes how the offerings of a firm meet customer needs and desires and thus create value for them. Once the decision factors underlying the customer needs have been described, the firm's offerings and value proposition characteristics must be identified and related to the different decision factor categories. Each characteristic is classified, depending on how it contributes to the firm's competitive positioning.

## 12.2.1  Competing on Differentiation or Uniqueness

A firm which exhibits its competitive advantage through differentiation, has unique traits in one or more elements related to its strategic focus, their relationships with the offerings elements OVP and OPS, and/or the external environment. Characteristics of the value proposition identified as unique are those that no other firm is currently offering and that customers are valuing. Uniqueness may result from specific capabilities, unique technologies, access to resources, or patents, to name just a few. Uniqueness is the most compelling attribute when identifying competitive advantages. These differentiations, either explicitly or implicitly visible, have an impact on the customers' decision journey. It is important to take a customer perspective when defining differentiation based competitive advantages. Unless customers see value for them from the differentiation traits, they provide no competitive advantage. Innovative firms typically compete through exhibiting a differentiating competitive advantage.

> **Example** Apple's AirPod headphones, combined with the Apple Watch, provide a unique way to place phone calls, that is not currently available from any competitor. Indeed, not having to grab a mobile phone to receive a call is unique and valued by customers whose job-to-be-done is answering phone calls in a hands-free and uncluttered (cable-less) environment.

Note that uniqueness must always relate to a specific customer need. Different customers have different needs, and thus may or may not value unique characteristics. Successful uniqueness characteristics are hard to copy by competitors and are preferred by customers over substitutes. In most cases, uniqueness is a temporary attribute. Its potential expiry must be dealt with as part of defining a firm's competitive positioning strategy.

> **Example** A typical example is Boing differentiating through focusing on twin-jet airplanes designed in direct collaboration with its customers, adding value for its customers by optimally addressing their jobs-to-be-done, in addition to reducing fuel costs when compared to four-engine airplanes of similar size and range.

### 12.2.2  Competing by Being Superior

Although superiority may be seen as a special case of uniqueness, superiority based competitive advantages focus on differentiating through the performance of business model element characteristics, rather than the characteristics themselves. Some of the value proposition characteristics may not be unique, but superior to those of competitors. If these characteristics are valued by customers, and thus have a positive impact on their decision process, they contribute to the firm's competitive positioning. Offering superior product or service quality is a typical superiority value proposition characteristic. Other superiority characteristics are ease of use, choice, after-sales-support, or being the cheapest. Superiority characteristics may also be related to emotional decision factors, like brand recognition. In contrast to uniqueness characteristics, superiority ones are easier to copy and compete against. As with uniqueness characteristics, superiority as a competitive advantage is specific to customer needs and desires. In the context of building a competitive advantage through superiority, firms need to find the right trade-off between value delivered to customers through superiority and the cost of achieving that superiority. Being superior at all cost is a failing strategy. For example, a digital watch being failsafe over a ten-year period, may be a superior characteristic, but due to the speed of technological advancement, not one that is valued by customers. Superiority based competitive advantages are often found in strategies focusing on commodity offerings with little opportunity to differentiation.

> **Example** A typical superiority strategy is competing on price, that is, being better at offering the lowest price for a specific offering aiming at getting an identical job of the customer done. This could be for example, offering the cheapest mobile phone subscription including unlimited data usage. Another superiority competitive advantage for a mobile phone operator may be offering the fastest possible internet connection in any location.

### 12.2.3  Handling Indifference

Most characteristics of the value proposition do not offer any differentiation, although they are necessary to satisfy the customer's jobs-to-be-done. They can be classified into the indifferent category. Indifferent characteristics are necessary, but do not add value that customers are willing to pay a premium for. They are as such not relevant for defining a firm's competitive advantage. They are called hygiene factors.

> **Example** Consider a bank offering a checking account. Being able to withdraw cash is considered an indifferent value proposition characteristic. It is required to satisfy the customer's need for cash. Customers may even be willing to pay for cash withdrawals, but the sole fact of offering access to cash is not influencing the customer's decision, and as such does not contribute to the firm's competitive positioning.

A firm must decide which characteristics of its value proposition to compete on and which to consider indifferent but necessary. Trying to compete on all characteristics of the value proposition will typically lead to failure. The competitive positioning of a firm is significantly defined by that decision. It should be distinct from that of its competitors.

## 12.3 The Competing Process

Defining a successful competitive advantage which is sustainable over time can be achieved by applying process G. The goal of process G is twofold. First, it aims at identifying the competitive advantage of the firm in the context of its detailed business model, eventually adjusting it. Second, it ensures that the competitive advantage can be sustainable by performing a game-theoretic analysis developing possible competitive strategy game plans, for reacting to external threats.

> **Process G—Defining a Sustainable Competitive Advantage Using Game Theory**
>
> G.1  Understanding the competitive landscape by
>
> - identifying key players, and
> - recognizing possible competition strategies applicable in the targeted industry
>
> G.2  Putting the designed business model into perspective by answering Porter's five questions on good strategy
> G.3  Determining the firm's competitive advantages centering in on its strategic focus
> G.4  Ensuring the sustainability of the competitive advantages in a dynamic environment using game theory by
>
> - identifying possible equilibria, and/or
> - developing and validating competitive strategy game plans

When identifying a sustainable competitive advantage fails, the strategy design process iterates back to the business model layer to address the identified issues. If the probability of the identified issues materializing is small enough, the firm may decide to accept certain reactions from other players without mitigating them. In this case, potential negative implications are documented as part of the strategy. Firms filing 10-K[1] or similar reports, are required to document these insights in the risk factors section.

## 12.4  The Competitive Landscape

Understanding the competitive landscape starts by identifying key players involved in the industry in which the firm aims at competing. Strategies on how to compete differ based on the industry and the structure of its participants. The competitive landscape analysis step G.1 addresses both.

### 12.4.1  Identifying Key Players

Building on Brandenburger and Nalebuff's (1995) company value net framework, seven categories of players whose actions may have a material impact on the success of the firm's strategy can be identified.

These are:

(1) *Customers*, both end-users and decision takers, as well as targeted non-customers.
(2) *Competitors*, including those that offer substitute products and services.
(3) *Complementors*, supporting the firm's offering to deliver value to its own customers in a complementary way.
(4) *Suppliers* of raw material and unfinished parts.
(5) *Employees*, especially those performing differentiating activities or participating in creating superiority.
(6) *Investors*, providing the necessary capital to implement the strategy.
(7) *Regulators*, ensuring fair behavior of all actors.

They are shown in Fig. 12.1. Although this list exhibits a significant resemblance with Porter's five-forces framework (Porter 1979), the competitive landscape analysis takes a confirmatory approach, rather than a designing one. This allows for a more open-minded design of the strategy than would be possible by building upon a five-forces analysis. The competitive environment analysis only addresses those players that are actual threats or opportunities to the firm and its strategy, rather than analyzing all potential players.

#### 12.4.1.1  Customers
Probably the most important player is the customer. A key question to answer is "what would make a customer change supplier/vendor?" The detailed business

---

[1]A 10-K form is an annual report required by the U.S. Securities and Exchange Commission (SEC), providing a comprehensive summary of a firm's financial characteristics.

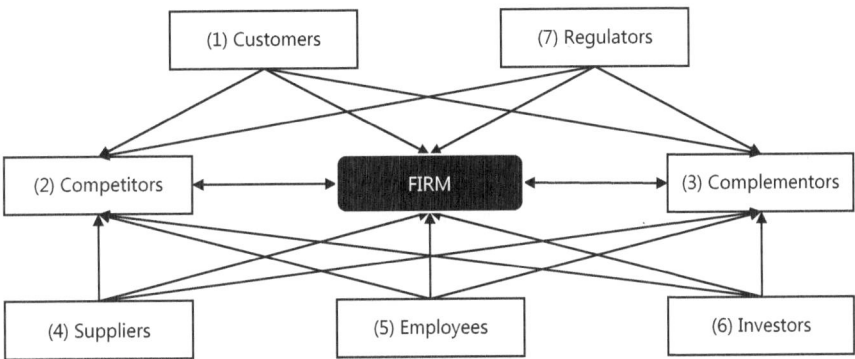

**Fig. 12.1** Key players affecting the success of the firm in a competitive environment

model, especially its customer relationship element (CR), provides a starting point for understanding the customer in the context of competition.

Which features, or lack of features, would make a customer become a non-customer? What role does the quality of the offering play in the customer's purchasing decision? How sensitive are customers to support services? What role do comments from other customers, for example on social media platforms, play in the customer's decision journey? How often could an offering break or fail, before a customer decides to switch supplier? What change in price, all else remaining the same, would make a customer look for a different offering? Answering those and similar questions allows defining the boundaries within which a customer feels valued.

**Example** Figure 12.2 illustrates the *customer value zone* concept related to the two dimensions processor speed and laptop price, for a computer manufacturer. As long as the laptop offerings of the firm remain within the value zone, the customer will not seek-out a

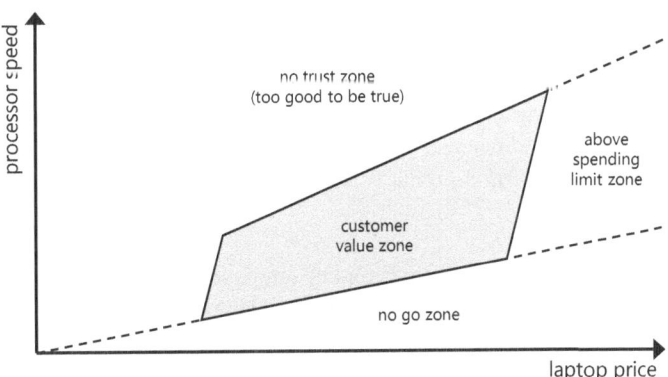

**Fig. 12.2** Value zone of a targeted customer segment related to laptop computers and the two variables processor speed and laptop price

different manufacturer. As such, competitive advantages must ensure that the offerings will remain in the customer value zone. Different competitive advantages define distinct customer value zones.

### 12.4.1.2  Competitors

Understanding competitors requires understanding if and how they might react to the firm's new or adjusted strategy, assuming unchanged customer jobs-to-be-done. If competitors decide to react, they often do so along any of the four dimensions, that is,

– *improving the perceived quality* of the offerings without charging for the improvements, to maintain or increase market share,
– *adapting the characteristics or features* of their offerings, including introducing new bundling, to attract customers specifically targeted by the firm's strategy,
– *offering superior support service*, from marketing, through sales, up to after-sales support, focusing on strengthening the customer relationship, or
– *reducing price*, to retain existing customers.

Any competitive action aims at changing the perceived value of the offering, as identified by the customers, with the goal to retain existing customers and/or attract new ones from competitors.

**Example** Consider for example Apple and its competitor Samsung. Apple introduced ApplePay in October 2014. Shortly thereafter, in August 2015, Samsung reacted by introducing a payment solution on its own, SamsungPay, to avoid losing customers that value the payment functionality to Apple and incentivize Apple customers interested in using their phones as mobile credit cards to switch vendor.

### 12.4.1.3  Complementors

Complementors are on often forgotten players in the competitive analysis. Complementors offer products or services that only add value to customers in conjunction with the firm's offering. From a customer perspective, complementors add the firm to the consideration set of potential customers, valuing both the offerings from the firm and its complementors. Successful complementors create win-win situations. But they may also introduce dependencies that the firm needs to monitor and potentially actively manage, as changes to the complementor's offerings may affect the value of the firm's products and services.

### 12.4.1.4  Employees

The success of any company depends on key employees, more precisely, their skills and relationships with customers. Any strategy defining its competitive advantage through key employees needs to understand what could make these employees leave. Typical employee criteria to consider at the strategy level are, in alphabetical order, appreciated, challenged, empowered, involved, mentored, trusted, valued, and well paid. Depending on the designed detailed business model, not only the skills of specific employees may be relevant, but also their availability. This is typically the case for consulting firms.

### 12.4.1.5 Suppliers
Although suppliers usually operate in a competitive environment, high value-adding supplies with significant bargaining power may decide to offer a firm exclusivity on certain raw materials or supplied parts or not work with a given firm. Such decisions by key suppliers may have an impact on the viability of the designed detailed business model and can impact the firm's competitive advantage. Therefore, any competitive landscape analysis needs to identify

– key *suppliers*, and
– their *bargaining power*,

and design activities to leverage opportunities and counter potential threats from them, to ensure sustainability of competitive advantages relying on suppliers. Typically, these may be long-term price agreements, guaranteed quantity availability, or exclusivity deals.

### 12.4.1.6 Investors
Some strategies require significant capital to grow (for example, to acquire new customers) and/or to operate (for example, to finance production equipment). Having access to investors satisfying these capital requirements provides a competitive advantage. As with other players, investors do not operate in a vacuum. They operate in a competitive landscape and have scarce capital to invest. The key competitive landscape question to answer, with respect to investors, is: "Under what circumstances would investors switch and invest in a competing firm?"

### 12.4.1.7 Regulators
The last but not the least important player to understand is the regulator. The term regulator is used as a synonym for governments, unions, and similar market force regulating actors. A sound competitive landscape analysis identifies all regulators, whose actions may impact the firm's strategy. Especially in highly regulated markets, like financial services, but also perceived less regulated markets, like taxi driving, the strategy needs to address legal and regulatory requirements upfront. A competitive advantage may be designed based on specific regulations or their interpretations.

### 12.4.2  Possible Strategies for Competing

There exist a number of generic, as well as industry specific, strategies for competing. They describe how typical firms behave in a given competitive environment. Competition strategies are rarely used on their own. They are applied to find a competitive equilibrium or to compete until there is a final winner.

*Signaling strategies* Rather than act, the firm signals to the market players, either explicitly or implicitly, how it would react to a given threat. If the signaled reaction is trustworthy, the threatened players may refrain from acting. Signaling is a low-cost strategy which works well in markets with a small number of trusted players. The challenge with signaling strategies is that, if signals are ignored, the signaling firm must react to avoid losing credibility.

*Monopolistic strategies* If the firm positions itself such as to be perceived as a quasi-monopolist, it may nip in the bud every potentially threatening player through signaling power. Consequently, players avoid competing with a monopolist strategy firm. This approach works well if the monopolist strategy firm can show enough power. It is typical in winner-takes-it-all type of industries, that is, industries primarily driven by size. Consider social media firms like Facebook or Twitter as typical firms following a monopolistic strategy.

*Capacity constraint strategies* Some firms operate in industries where capacities are constrained, usually due to limited availability of raw materials or adequately skilled human resources. If, in addition, large fixed costs or investments are a precondition for competing, players may aim at producing at a capacity that would make any other firm entering the market operate at a loss. Rather than implement a monopolistic strategy, firms implementing a capacity constrained strategy often only need 20–30% of market share, depending on the surrounding parameters, to succeed and deter new entrants. The steel industry is typical industry in which a capacity constraint strategy can work.

*Cannibalization or market squeezing strategies* Firms aiming at competing through cannibalization offer products and services that address similar jobs-to-be-done at a discount price with the sole goal to push other firms out of the market by making them unprofitable. Once the other players have exited the market, the cannibalizing firm increases prices again to recoup the suffered losses. Cannibalization strategies often target firms streamlining their portfolio of offerings. They work well in low-margin industries. They often require significant up-front capital.

*Price elasticity strategies* Competing on price elasticity aims at outperforming competitors by better understanding the price elasticity and attracting new customers at the margin. Price elasticity strategies are tightly related to a superior understanding of the willingness to pay of customers and the willingness to sell of suppliers. The can be very successful and hard to imitate, if implemented well.

When studying possible competition strategies, it is important to also understand how customers will react. Customers may wait before they switch the firm they are buying from. Customers may interpret prices, and especially price changes differently than do firms. They may try to game the system by anticipating competitive reactions. Customer reactions or failures to react in an expected way may have a significant impact on the success of implementing any competition strategy.

## 12.5  The Business Model in the Competitive Environment

No name is more closely related to the concept of competitive advantage in strategy than Porter (1980, 1985). According to his line of thoughts, strategy is about choice, namely choosing who to serve and who not to serve, what to do and what not to do, resulting in a unique way on how to compete. Strategy is the antidote to competition. The detailed business model provides one perspective on the firm's competitive positioning. Competitive analysis aims at ensuring that the firm's strategy, including its detailed business model, offers a unique way to be superior and/or different from competitors.

In 1996, Porter published a paper in the Harvard Business Review called "What is strategy?" (Porter 1996) summarizing what characterizes a good strategy. Any sound strategy providing a competitive advantage is based on business model characteristics resulting from answering five key questions (Magretta 2012):

(1) *What distinguishes the value proposition of the firm from that of competitors?* Answering this question requires understanding which customers to serve and which not to serve. It also means defining which customer jobs-to-be-done to satisfy and which not. It means showing how value is created for customers that results in profitability for the firm.
(2) *Which activities does the firm perform in a different or superior way than its competitors? What is the uniqueness of the firm's value chain?* These questions are answered by taking an inward viewpoint and focusing on understanding how the tailored or unique elements of the firm's value chain support delivering the value proposition. The identified activities form the firm's core competencies.
(3) *Which trade-offs, different from those of its competitors, does the firm make?* Strategy is about choice. Choice requires trade-offs. Identifying trade-offs allows understanding how the firm creates a sustainable competitive advantage. It also means clearly defining what the firm does not offer, who the firm is not serving, and where the firm is not competing.
(4) *Which strategic fits does the firm amplify?* Strategic fit means relating individual activities of the value chain to each other, leveraging core competencies to create value in excess of that of the individual activities in a way that is difficult, if not impossible, to imitate.

(5) *How is the strategy supporting continuity over time?* Even though strategy is about change, continuity over time of key elements of the strategy is an integral property for achieving a sustainable competitive advantage. Continuity reinforces identity and trust. It also helps building differentiation though lasting relationships with customers, partners, and suppliers.

Identifying the firm's competitive advantages and ensuring that they are not transient requires relating the answers to these five questions to the elements of the firm's detailed business model. Depending on the answers given, the detailed business model may be iteratively refined or amended.

## 12.6  Designing the Firm's Competitive Advantage

Designing the firm's competitive advantage requires answering the key question:

*Why should a customer buy the firm's offering*
*rather than that of its competitors?*

Answering that question can be subdivided into answering three related questions:

(1)  What makes the detailed business model of the firm different from or superior to that of competitors?
(2)  Why is the identified differentiation or superiority preferred and valued by the targeted customers?
(3)  How can the identified differentiation or superiority be sustained over time?

First, insights are gained from the answers to Porter's five key questions about strategy. Objectivity is important. There is no value in fooling oneself. The often-heard argument "we have the best employees" does not provide a competitive advantage unless "best" is valued by customers as distinct or superior.

Next, the answers to Porter's questions are related to the different elements of the detailed business model. The detailed business model elements are re-assessed and potentially refined, considering the competitive landscape and its players. Each element is reviewed in the context of it offering differentiation or superiority when compared to competitors' business models. The competitive advantages identified should be distinct, or at least sufficiently different, from the ones of competitors. They need to be well articulated and understood by the target customer segments to ensure they act on them. They should be hard to imitate and/or exhibit little interest in copying. A firm should limit its competitive advantages to a small number. The quality and sustainability of competitive advantages are more important than their quantity.

### 12.6.1 Customers Based Competitive Advantage

Customer centric competitive advantages are identified by reviewing the customer related elements of the firm's detailed business model, that is, the CR and CD elements. They may also be found by understanding relationships between customer elements CS and CJ, and the offerings elements OVP and OPS. The third area leading to identifying competitive advantages are links between the detailed business model and the external environment. A competitive advantage may be identified by focusing on underserved customer segments or addressing previously unmet jobs-to-be-done. Capabilities allowing to understand the specificities of customer jobs-to-be-done, can also be translated into a competitive advantage, especially when combined with customizable offerings.

> **Example** In its early days Research in Motion (RIM), the provider of the legendary Blackberry phones, defined its competitive advantage by targeting business customers and their job-to-be-done of secure communication, while competitors targeted private customers and corporations focused on buying on price rather than on specific features.

A competitive advantage can also be identified as the capability of retaining customers (CR element) and spurring recurring purchases (CJ element), by introducing switching costs.

> **Example** Nestle's Nespresso gained a competitive advantage by introduce switching costs through patenting their coffee capsule design.

> **Example** For many firms, like Starbucks or Nike, their brand is a hard to imitate competitive advantage.

Other areas where competitive advantages can be designed into the detailed business model are around delivering approaches (CD element), by better understanding where and when to deliver purchased products and services.

The *Competitive Positioning Canvas*[2] (*CPC*), shown in Fig. 12.3, is a framework to document insights and knowledge that support identifying a firm's competitive advantage focusing on customers and their jobs-to-be-done. The CPC is not the firm's competitive advantage by itself but a tool that provides a common language to executives, strategists, and consultants, for leading the discussion and decision about competitive positioning. It helps take a different perspective and ensures that no key insights are missed.

Given one or a group of customer needs and jobs-to-be-done, the CPC allows identifying how customers define value in their utility function. It first focuses on *rational decision factors*, meaning understanding what are the must have and the nice to have *value* characteristics driving customer decisions. These are typically required features, like product and service quality, usability, or after-sales support, to name just a few. It also means understanding the customers' perception of *costs*,

---

[2]The *Competitive Positioning Canvas* builds upon an INSIGHT published by innovate.d llc in January 2019 as "Understanding a firm's competitive positioning". It can be found under https://www.innovate-d.com/insight-101/.

| Customer jobs-to-be-done | Rational decision factors | | Emotional decision factors |
|---|---|---|---|
| Job(s) to be done | Value | Costs | |
| **Needs and desires derived from customer jobs-to-be-done** | **Must have & nice to have**<br>• Features    • Availability<br>• Ease of use/    • Uniqueness<br>   sophistication    • Custom<br>• Configurability    made<br>• Support<br>   service<br>• Offerings<br>   quality | **Price & cost of access**<br>• Cheapest    • Price<br>• Value for     transparency<br>   money    • Finding costs<br>• Competitive    • Cost of<br>   price     access<br>• Premium    • Due diligence<br>   price     costs<br>• Perceived    • Reputation<br>   free goodies     value | • Personaliza    • Fair price<br>   tion<br>• Choice<br>• Trust<br>• Status<br>• Popularity<br>• Reputation<br>• Uniqueness |

| Offerings | Rational decision factors | | Emotional decision factors |
|---|---|---|---|
| Product / Service | Value | Costs | |
| **Unique** | • Capabilities<br>• Technologies<br>• Know-how and experience<br>• Access to natural resources<br>• Patents and intellectual property<br>• Features<br>• Novelty | • Pricing model<br>• Bundling<br>• Distribution network<br>• Timing of payments<br>• Units of value | • Perceived innovativeness<br>• Access to customers<br>• Distributors<br>• Perceived offering status, e.g.,<br>   premium<br>• Opinion leaders/influences<br>• Brand value |
| **Superior** | • Quality of offering<br>• After-sales support<br>• Skills of labor<br>• Availability and quality of<br>   natural resources<br>• Quantity and quality of features | • Price level<br>• Search costs to find offerings<br>• Due diligence costs | • Brand value<br>• User reviews<br>• Trust factors<br>• Offerings shelf size |
| **Indifferent** | • Easily replicable traits<br>• Traits with a competitive<br>   disadvantage<br>• Functionalities required to get<br>   the job done but not explicitly<br>   valued by customers | • Easily replicable traits<br>• Traits with a competitive<br>   disadvantage<br>• Functionalities required to get<br>   the job done without explicit<br>   customer price sensitivity | • Easily replicable traits<br>• Traits with a competitive<br>   disadvantage<br>• Must have functional<br>   requirements without any<br>   emotional importance/relevance |

(Value proposition / Offering)

**Fig. 12.3** The Competitive Position Canvas (CPC) providing a common language for describing the characteristics that allow a firm to describe its competitive advantage

looking a price (cheapest, value for money, competitive, premium) as well as access costs (costs related to searching for an offering and buying it).

The second dimension to explore to understand customers decision factors is the *emotional dimension*. Emotional decision factors can be classified based on the nature of the relationship between the firm and the customers, that is, either one-way (brand, reputation, advertising) or bi-directional (customer intimacy, pro-activeness, distribution channels).

As shown in Fig. 12.3, the top part of the CPC represents the considered jobs-to-be-done and relates them to the needs and desires derived from the jobs the customer wants to get done. In a second step, the bottom part of the CPC represents the offerings characteristics and documents the value proposition characteristics by classifying them into the three possible competitive advantage categories, that is, uniqueness, superiority, or indifference. Finally, the value proposition elements are matched to the customer decision factors, ensuring optimal competitive advantage by relating the top part of the CPC to the bottom one.

## 12.6.2 Offerings Based Competitive Advantage

The most common area where competitive advantages are found when focusing on an offerings strategic focus is in the products and services element (OPS), as well as the associated value proposition element (OVP) of the detailed business model. Typically, hard to copy features lead to differentiation. Superior quality based on unique production and quality control processes are another area where a firm can generate a superiority based competitive advantage. Competitive advantages do not have to directly relate to the core of the offering. They may be based on support services or even packaging of the products offered. Consider a premium airline differentiating through on-board service, rather than flight schedules. The CPC in Fig. 12.3 helps identify offering based competitive advantages, starting with the bottom part, the offerings part, and relating them to the top part, the jobs-to-be-done part, in a second step.

## 12.6.3 Capabilities Based Competitive Advantage

A firm exhibits capability based competitive advantages by having unique capabilities, for example, production machines, physical resources, processes, intellectual property, or patents. Capability based competitive advantages are primarily designed around economies of scale, providing superiority, and economies of scope, providing differentiation. Competitive advantage can be developed through combining existing capabilities in a unique way along Porter's line of amplifying strategic fits. A competitive advantage can also be achieved by leveraging skills in a way hard to imitate. Gaining efficiency through outsourcing and managing the relationships with partners and suppliers can also lead to a competitive advantage, assuming that part of the underlying value can be made available to the customers.

## 12.6.4 Financials Based Competitive Advantage

Many firms define their competitive advantage through being able to match any competitor's price. Although challenging, due its transient nature, and the risk of being cornered or squeezed-out of the market by larger competitors, competing on price can be a possible competitive advantage. Firms focusing on price-based competitive advantages are often found in industries that are perceived as offering commodity products or services with little or no differentiation, like consumer electronics, the airline industry, or grocery stores.

More recently, competitive advantages build around unique pricing models have emerged. Rather than taking a firm-centric approach to pricing, competitive pricing models are designed around understanding how customers perceive paying for the value delivered by a product or service.

Another way to achieve a financials based competitive advantage is coming up with a unique way of dealing with price externalities, for example, forging exclusive agreements with perishable resources suppliers.

## 12.7   Winning the Competition Game by Sustaining a Competitive Advantage Using Game Theory

Defining and implementing a competitive advantage often results in adverse reactions from competitors that need to be countered to win the competition game and remain profitable. Winning the competition game means being prepared and having though-through scenarios for all major competitive reactions. When designing potential actions to react to competitive threats, alternative approaches to competition need to be identified.

Consider a firm that competes on differentiation, through patented features. There exists a threat from competitors adding features to their offerings that substitute the value provided by the patented features without infringing on any patents. One way of addressing such a threat is through adjusting the strategy by re-defining the target customer segment such that the competitor's substitute is no longer considered a viable alternative. Another way of addressing that threat is inventing new features valued higher by customers than substitute features offered by competitors. Another alternative would be improving upon the existing patented features by showing their superiority to the substitutes from competitors. A fourth alternative would be competing on price, discounting the patented offering and providing a superior value/cost ratio to customers.

> **Example** A typical example of regaining competitive advantage through unique services models has been implemented by Lenovo, the computer manufacturer. It services computers at the buyer's location worldwide (or nearly), rather than having customers send-in their broken computers for repair.

Examples of distinct pricing models are pay-as-you-go models, no longer needing up-front payments or introducing in-app purchase options that tie the price more closely to the value delivered by a specific feature.

Some of these examples may seem obvious, some far-fetched. The one thing they all have in common, is that they are based on creative ideas designed, validated, and implemented, focusing on offering value to customers in a competitive way.

Competition can be described as a game with two or more players (Morgenstern and von Neumann 1947; Nash 1950, 1951; Dresher 1961; Ghemawat 1997; Dixit and Nalebuff 2008). In some cases, the game is a zero-sum game with a winner and a loser, like chess or checkers. Most games modeling economic situations, are non-cooperative games. The typical competition game is based on imperfect information and includes some degree of randomness.

Game theory provides frameworks for studying competing strategy games and their impact on competitive actions of the players. Their use in business is still in early stages (Brandenburger and Nalebuff 1995; Ghemawat 1997). There exist two types of game theories that fit well into the abductive design thinking-based strategy design process. They are

- *equilibrium theories*, like the Nash equilibrium, allowing to study and understand competition through *differentiation*, and
- *game tree theories*, like the min-max approach, focusing on determining the optimal action to take under uncertainty when competing through *superiority*, assuming that competitors aim at maximizing their utility in a rational way.

Rather than start an extensive analysis of a firm's strategy using game theory, I recommend putting the focus on those aspects of game theory that help validate or invalidate the effectiveness of the designed strategy to competitive threats. Only reviewing a small subset of options is needed to understand potential competition and designing possible scenarios using game theory. Game theory helps analyze the competitive environment by supporting the validation of the designed strategy, especially focusing on identifying potential flaws and being prepared for competitive reactions. Game theoretical analysis in strategy is about being prepared to play the competitive game under uncertainty.

## 12.7.1 Competitive Equilibrium

Understanding possible competitive equilibria is based on both the firm and the competitors choosing to compete on being different. For example, a firm and its competitor (assuming for the sake of simplicity only one competitor) may have the choice to either focus on private or on corporate customers, as illustrated in Fig. 12.4. Game theory would require determining the value of each of the four options. This is sound in theory, but much harder in practice. And it gets even harder if considering more than one competitor. Therefore, focusing on qualitative assertions, provides possible choices. Once choices are characterized, an equilibrium state is sought, as shown in Fig. 12.4a. An equilibrium state is a state where both firms are better off than any other alternative state. The focus is on both players, the firm and its competitor, rather than one player alone. In some situations, as illustrated in Fig. 12.4b, there does not exist an equilibrium situation, requiring alternative competitive analysis to design possible scenarios to win the competitive game.

**(b)** 

| firm having similar capabilities and cost structures as the competitor | competitor having similar capabilities and cost structures as the firm | |
|---|---|---|
| | corporate customers | private customers |
| **corporate customers** | both firm and competitor compete on price as there is no other differentiation possible, resulting in a price war<br>firm loses<br>competitor loses | although focusing on private customers, the competitor outcompetes because economies of scale allow for cheaper prices, also attracting corporate customers<br>firm loses<br>competitor wins |
| **private customers** | although focusing on private customers, the firm outcompetes because economies of scale allow for cheaper prices, also attracting corporate customers<br>firm wins<br>competitor loses | both firm and competitor compete on price as there is no other differentiation, resulting in a price war<br>firm loses<br>competitor loses |

**(b) Situation where no equilibrium exists when competing around customer segments served**

**(a)**

| firm with superior capabilities to serve private customers in an effective way | competitor with superior service capabilities for corporate customers | |
|---|---|---|
| | corporate customers | private customers |
| **corporate customers** | all corporate customers switch to the competitor because of its superior capabilities to service corporate customers<br>firm loses<br>competitor wins | some corporate customers switch, but are unhappy because firm fails to offer expected service<br>private customers switch to competitor<br>firm marginally loses<br>competitor wins |
| **private customers** | private customers switch to the firm<br>corporate customers switch to competitor who already has an advantage in serving them<br>firm wins<br>competitor wins | competition for the same customer segment results in a price war, as private customers primarily buy on price<br>firm loses<br>competitor loses |

**(a) Equilibrium when competing around customer segments served**

**Fig. 12.4** Illustration of two players focusing their competitive advantage on either servicing corporate or private customers

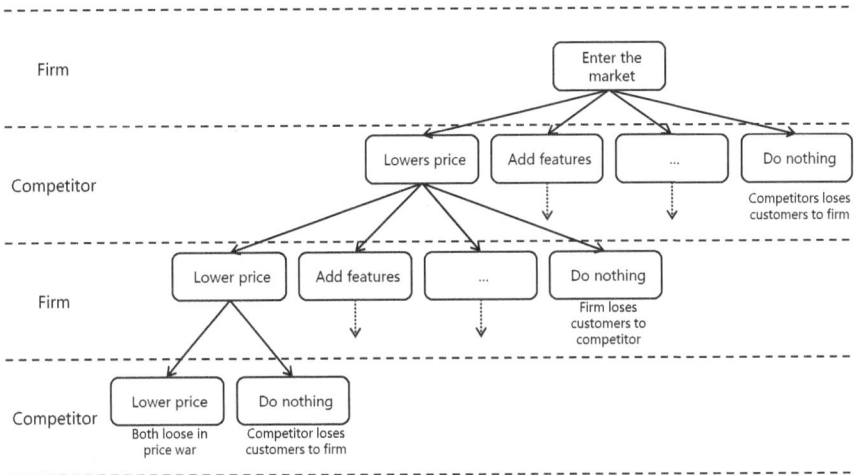

**Fig. 12.5** Subset of a sample game tree modeling competition between two low-cost mobile phone manufacturers

Equilibria are often temporary in nature, as the value of the different options changes over time. Real life is usually more complex than the examples in Fig. 12.4, and includes more than two players and more than two options. Approximations and the application of common sense during the analysis and modeling is needed to achieve meaningful results in reasonable time. In most cases where players compete on differentiation, the equilibrium analysis is key to avoid leaving money on the table.

## 12.7.2 Modeling Competition Using Game Trees

Game tree theory, also called min-max theory, takes a different approach than equilibrium theory. Rather than looking for an equilibrium, it models actions and reactions of the involved players over time to find the most promising decisions, similar to how chess is plaid.

> **Example** To illustrate the modeling tool, consider two payers offering similar low-end mobile phones. At any given point in time, each play has three options, that is (i) reduce the price, (ii) add new features, or (iii) do nothing. Figure 12.5 illustrates a subset of the possible decisions each company can take represented by a decision or game tree. Companies alternatively decide about their next move up to the point where one either loses, wins, or both are stuck in a draw situation. Such a situation is called a leaf in the game tree. Once the game tree has been constructed, the value of each intermediary node is determined, assuming that each player always choses the move that leads to the best outcome from its perspective.

As in the equilibrium approach, using game tree theory requires common sense, especially to value the quality of a decision at a given leaf of the game tree. The approach helps strategy designers think though multiple options. It reduces the risk being caught by surprise when competitors react to possible threats.

## References

Brandenburger, A. M., & Nalebuff, B. J. (1995). The right game: Use game theory to shape strategy. *Harvard Business Review, 76*(7), 57–71.

Christensen, C. M., Hall, T., Dillon, K., & Duncan, D. S. (2016). *Competing against luck: The story of innovation and customer choice.* New York, NY: HarperCollins Publishers.

Dixit, A. K., & Nalebuff, B. J. (2008). *The art of strategy: A game theorist's guide to success in business and life.* New York, NY: W. W. Norton & Company.

Dresher, M. (1961). *Games of strategy: Theory and applications.* Upper Saddle River, NJ: Prentice Hall.

Ghemawat, P. (1997). *Games businesses play: Cases and models.* Cambridge, MA: MIT Press.

Magretta, J. (2012). *Understanding Micahel Porter.* Boston, MA: Harvard Business Review Press.

Morgenstern, O., & von Neumann, J. (1947). The theory of games and economic behavior. Princeton, NJ: Princeton University Press.

Nash, J. F. (1950). Equilibrium points in N-person games. *Proceedings of the National Academy of Sciences of the United States of America, 36*(1), 48–49.

Nash, J. F. (1951). Non-cooperative games. *Annals of Mathematics, 54*(2), 286–295.

Porter, M. E. (1979). How competitive forces shape strategy. *Harvard Business Review, 57*(2), 137–145.

Porter, M. E. (1980). *Competitive strategy.* New York, NY: The Free Press.

Porter, M. E. (1985). *Competitive advantage.* New York, NY: The Free Press.

Porter, M. E. (1996). What is strategy? *Harvard Business Review, 74*(6), 61–78.

# Laying the Groundwork for Strategy Implementation Through Stakeholder Focused Communication

*The single biggest problem in communication is the illusion that it has taken place*—George Bernard Shaw

The last but not the least step of the strategy design process is communicating the designed strategy and the associated detailed business model to stakeholders, selling it to the crowd, so to speak. Communication has multiple goals that must be addressed to be successful (Jones 2008).

First, communication is about *informing*. Only if you know where to go, can you find the way to get there. This means that the strategy message must be customized for the target audience. The information should focus on the goal, rather than the path to achieve it. It is important to choose the appropriate level of abstraction.

Second, communicating strategy is about *setting the stage for change*. This means, ensuring that the stakeholders understand the new or revised strategy, so that they can identify and plan required changes within their area of influence. Communicating the strategy is the first time where the play meets the audience. As a proverb says, you never get a second chance to make a first impression. Strategy is about making choices, that is, which game to play, how to win, and how to be successful. It is also about what not to do, whom not to serve, and what not to offer. All these elements need to be addressed in the messaging.

Third, communicating strategy is about *convincing* the stakeholders that the strategy choices made are optimal for the firm and support it to be profitable in a sustainable way. Depending on the perspective, communicating strategy is also about *reassuring* people.

Fourth, any strategy is only as successful as the individuals implementing it. A successfully communicating strategy requires *engaging* people. This means, capturing the audience both rationally and emotionally. At the end, people must embrace the strategy.

Fifth and final, communicating strategy can only be successful if the recipients of the messages feel *be taken seriously*. The owners of the strategy must therefore

© Springer Nature Switzerland AG 2020
C. Diderich, *Design Thinking for Strategy*, Management for Professionals,
https://doi.org/10.1007/978-3-030-25875-7_13

stand up for it, show that they really believe in the success of it, and will do whatever it takes to make it happen.

## 13.1   The Communicating Process

It is important to follow a structured approach, based on the design thinking principles of iteratively designing and validating, to achieve the aforementioned five goals. Process K describes seven steps to follow for successfully communicating strategy.

**Process K—Communicating Strategy**

K.1   Understanding the ground rules
K.2   Identifying the audience/stakeholders (to whom to communicate)
K.3   Selecting the most appropriate communication channels
K.4   Laying-out the timeline (when to communicate)
K.5   Preparing the strategy message (what to communicate)
K.6   Designing how to tell the story (how to communicate)
K.7   Reviewing and ensuring that the strategy message is understood (validating the message)

Communicating strategy must ensure that the distinct aspects underlying success are addressed in a thoughtful and audience-centric way. As with all processes described in this book, individual stages may be lengthened or shortened depending on the needs of the specific firm. In the light of the iterative nature of design thinking, the different process steps may be, and often are, iterated to maximize their value. This is especially true for the steps K.5–K.7.

## 13.2   Understanding the Ground Rules

The ground rules of communicating strategy set the stage and define the foundation for success (Jones 2008). Ignoring one or more of these rules increases the probability of failure, that is, that the new or revised strategy is misunderstood, that stakeholders do not believe in the associated changes, or that senior management is not serious about the targeted way of successfully competing. Six key ground rules must be observed:

(1) *The CEO[1], as the owner of the strategy, must be the owner of the strategy message.* This ensures responsibility for the message and supports its seriousness.

(2) *Strategy communication must be done by real people.* Although there exist many tools to support communication, such as social media, newsletters, or videos, only people can deliver a convincing message. Stakeholders take the message delivered only as seriously as they take the messenger delivering it. Strategy communication is a top-down process that cannot be delegated or outsourced.

(3) *Successful strategy communication requires consistency in the message delivered by various sources.* Although the delivery of the message can and should be tailored to the target audience, the core of the strategy message must remain constant. The strategy message must be recognizable whoever communicates it. Consistency is key to convincing, reassuring, and showing seriousness. As such, anyone communicating about the strategy must have an optimal understanding of it.

(4) *Strategy communication must be integrated into the daily routine.* The biggest mistake to make is treating strategy communication as a one-off event. The underlying message must be repeated over and over again in a consistent way. Understanding the strategy should be as common as reading e-mails. The storyline used to communicate the strategy message can be adjusted over time, especially with respect to its depth. Stakeholders should dream about the strategy without having nightmares.

(5) *Communicating strategy must avoid surprises.* Stakeholders must get prepared over time for the details of the new or revised strategy. Best results are achieved by choosing a gradual approach to communication rather than a big bang set-up. People are only able to capture so much of a message at a time. Achieving buy-in is much easier if people can relate the strategy message delivered to what they already know, understand, and especially how it impacts them.

(6) Finally, *any strategy communication message should be tested for its impact before being delivered.* Getting the message content as well as the delivering platform right is hard. This is especially the case when the perception is nearly solely defined by the receiver of the message. Validating the strategy message with a small audience and adjusting it based on their feedback is important. The biggest challenge in testing the strategy communication message is ensuring appropriateness of the test audience combined with confidentiality of the message.

The decisions taken during the subsequent steps K.2–K.7 of the strategy communication process should be cross-checked against these six ground rules and adjusted, if and when deemed necessary.

---

[1]Depending on the legal framework, the chairman of the board of directors the board as a whole, the CEO, or the executive committee, may be the owner of the strategy. This chapter refers to the owner of the strategy as the CEO of the firm. It is important not to confound the designer of the strategy or the owner of the strategy design process, with the owner of the strategy in its entirety.

## 13.3  Identifying the Audience

Before being able to communicate the strategy message, the audience and their expectations must be identified. This is no different from preparing for a speech. The audience can be subdivided into two categories, that is,

- the *internal audience*, also called the active stakeholders, and
- the *external audience* or the passive stakeholders.

Identifying the audience is similar to identifying target populations. The clustering of stakeholders should be based on common needs in knowing and understanding the strategy message. This means, taking the viewpoint of the stakeholders and asking the question what they would need to know to ensure that the strategy message is understood. In addition, communicating to well-thought through clusters of stakeholders reinforces the strategy message.

### 13.3.1  Internal Audience

The internal audience, namely managers and employees, can be subdivided into three categories, that is,

- *managers*, asking the question "how can I get the strategy message through to my direct reports and employees?",
- *affected employees*, wanting to know what the new or revised strategy means for them in particular, and
- *not affected employees*, attempting to understand why they should care and how they should react.

Different sub-categories of affected employees requiring distinct strategy messages may be defined. The value chain framework provides a base to consider for sub-segmenting the affected employees audience, distinguishing, for example, between sales people, operations experts, and support staff. Distinguishing on seniority or title, although common, is sub-optimal.

Often, employees have two roles at the same time, being manager and affected or not affected employee. Depending on their role, the strategy message must be adapted.

In situations where the strategy to be communicated is completely new or disruptive, the category of not affected employees may not exist. If it exists, it is important not to ignore it as these employees have a role in the overall structure of the firm and may become supporters or saboteurs, depending on how they perceive strategy communication is handled. Perception is reality.

## 13.3.2  External Audience

It would be a big mistake to ignore the external audience when introducing a new or revising an existing strategy. Although external stakeholders are not directly affected by the resulting strategic changes, their reaction may be critical for success. The four most prominent external stakeholders are:

(1) *Investors*, trying to understand what is different and especially what is superior in the new or revised strategy based on their perception of the industry, why they should embrace the new or revised strategy and not divest.
(2) *Regulators* and unions, looking for potential challenges that could arise from their perspective and how they should cope with them.
(3) *The media*, looking for stories to tell, finding the grain of salt in the sauce.
(4) *Customers*, wanting to understand what will change from their perspective, what potential disruptions to expect, which habits they may have to change, or if they have to look for alternatives.

Depending on the industry and the characteristics of the strategy to be communicated, additional external stakeholders may be considered. The environmental analysis approach to stakeholder identification (Chap. 6), provides a good starting point for insights on which stakeholders to consider.

## 13.3.3  Looking at the Audience from a Different Perspective

The different stakeholders can be classified along their perceived role during strategy communication, rather than segmenting the audience based on organizational, hierarchical, or seniority aspects. Five key roles, which have an impact on the strategy message, can be identified:

(1) *Decision owners*, responsible for the outcome of the strategy and its implementation.
(2) *Implementers*, in charge of implementing the strategy and taking tactical and operational decisions based on guidelines from decision owners.
(3) *Networkers* or highly connected people, which can efficiently relate the strategy messages within the firm and to external stakeholders, independent of their functional or hierarchical roles.
(4) *Influencers*, usually not directly involved with the strategy implementation, but which are interested in leveraging and spreading a positive message.
(5) *Saboteurs*, having a self-interest not aligned with the new or revised strategy and therefore actively pursuing to undermine the success of the strategy.

Successful strategy communication depends on addressing and possibly leveraging these informal roles.

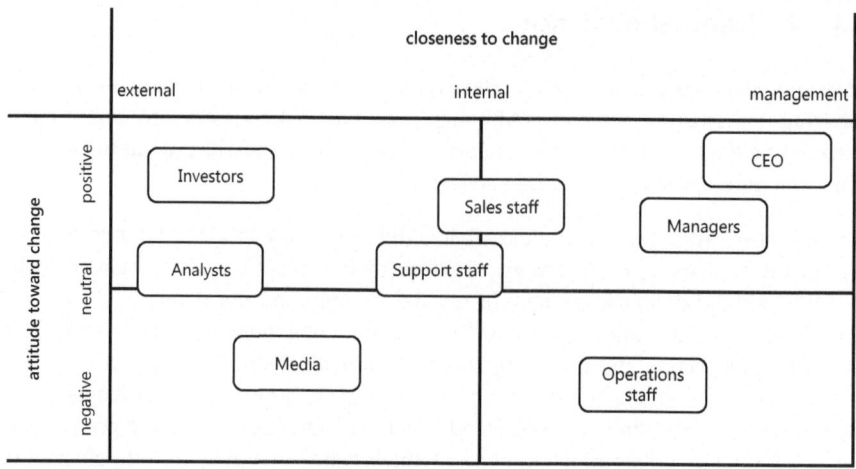

**Fig. 13.1** Sample subset of stakeholders to consider when communicating about a revised strategy

Different stakeholders are identified, regrouped, and classified based on their closeness to the strategy's success and their expected position towards change. Figure 13.1 illustrates a subset of a typical audience for communication about a revised strategy.

## 13.4 Selecting Communication Channels

There exist different channels which can be used to communicate about strategy. Each channel has its own advantages and drawbacks. A good communication strategy relies on combining multiple channels based on the target audience. Independent of which channel is used, the strategy message must be adapted to the specific channel. For example, when choosing a video message to be broadcasted to employees, it should be to the point, that is, like an elevator pitch and delivered by the CEO.

When choosing a given channel, it is important to use technology to support message delivery rather than replace the human aspects. Two-way communication, or at least including a feedback loop, is to be preferred to one-way communication, whenever possible.

## 13.4.1 Face-to-Face Communication

Face-to-face communication ensures that the strategy message is delivered in person. Typical forms are town hall meetings, on-stage interviews, or short presentations, followed by questions and answers sessions.

*Advantages* Face-to-face communication supports delivering an authentic message. It allows for two-way communication through interaction and real-time verbal and non-verbal feedback. The focus is on people rather than on anonymous media.

*Drawbacks* Depending on how the message is delivered, there exists the risk of diluting the message by going off-script. A right trade-off between on-script and authenticity must be found. Especially questions and answers sessions incur the risk of losing the core message, by focusing on answering specific questions. Availability of the audience and accessibility to the message deliverer are other challenges to overcome.

Any effective strategy communication requires including at least one face-to-face communication to each of the identified audience categories.

## 13.4.2 Electronic Communication

In the age of social media, electronic communication is the most prominent communication channel. Messages may be audio, video, illustrations, photos, or text, or a combination thereof. Typical channels are e-mail, messengers, such as WhatsApp, Signal, or WeChat, in-house blogs, chat forums, up to social media channels, such as Yammer, Facebook, Twitter, Snapchat, or even Instagram. The main challenge is selecting the right channel without overdoing it. It is important to understand under which conditions, that is, where and when, the audience will best ingest the strategy message.

*Advantages* The key advantage of electronic communication is its speed of delivery. In addition, due to the nature of the channel, it is much easier to control the message and tailor it to the audience. In the extreme case the message may be personalized for an audience of one. Different media, including visual and text, can be combined to optimally deliver the strategy message.

*Drawbacks* Electronic communication makes it harder to control if, when, and how, the strategy message is consumed. There exists the risk of the message not being understood or misinterpreted. No or only uncontrolled feedback is possible.

Electronic communication should be used when speed and control of the message is important. Unmoderated feedback, such as seen on social media, should be avoided. Electronic communication ensures that a large and de-centralized audience can be addressed rapidly. Often electronic communication precedes face-to-face communication. It is also recommended for summarizing the leanings from face-to-face meetings, making them available to a broad audience.

### 13.4.3   Print Communication

In the era before the internet, print was the primary communication channel. Typical print media are letters, in-house magazines, and newspaper articles.

*Advantages* Print communication allows for a physical delivery of the strategy message. It is a less transient medium than electronic communication. As with electronic communication, print media allow reaching a broad audience, especially when communicating to external stakeholders.

*Drawbacks* Print communication is, by its nature, slow to deliver. It offers limited space and is costlier that electronic media. As for electronic media, there exists a risk of misunderstanding of the message. Practically no feedback is possible. In addition, the message may be perceived as impersonal.

Print communication is best used to confirm electronic and/or face-to-face communication. It gives the strategy message a lasting touch. For example, summarizing the strategy on a one-page laminated print, may help employees remember it over time.

### 13.5   Laying-Out the Timeline

There exist two possible extreme approaches for when to communicate a new or modified strategy. These are,

- the *big bang approach*, where the whole strategy message is communicated to everyone at once, and
- the *iterative approach*, where elements of the strategy message are communicated in a just-in-time way.

In practice, a mix of both approaches should be relied upon. The primary constraint to be satisfied is that strategy communication must only consider what has been finally decided and approved. Communication must never be about strategy design work in progress. This does not mean, that it is forbidden to inform about ongoing strategy design work. But the messaging must be clearly separated. Undecided options should never be communicated outside the strategy design team, except as part of experiments and labeled as such.

Regulatory requirements often impose a big bang type of strategy communication approach, especially with respect to informing the external audience. The big bang approach allows showing the often-required sense of urgency associated with change in strategy. Challenges are keeping the message secret ahead of release and avoiding leaking of insights upfront.

Choosing an iterative approach allows incorporating feedback received into the message. It allows expanding the message over time and permits focusing on those aspects that matter most at a specific point in time.

Usually no more than two to four weeks should elapse between the first-time strategy is communicated and the first implementation steps are taken. Whichever option, or combination of options, is chosen, strategy communication is an ongoing process. It must become part of the firm's culture ensuring that individual actions are aligned with the overall strategy.

## 13.6  Preparing the Message

So far, numerous insights into the strategy of the firm have been developed. Key insight gained are

- the *industry* in which to compete,
- the *strategic focus* identifying the primary dimension along which to excel and differentiate,
- a description of the firm's *detailed business model*, describing targeted customer segments, value propositions, capabilities, and a financial plan, as well as how these components interact,
- the *competitive advantages* that will allow the firm to successfully compete, and
- *game plans* on how to react to moves by competitors.

Strategy is about choices. Communicating strategy is about making these choices transparent and describing how they lead to superior performance and value for all stakeholders. Preparing the strategy message is about content. Telling the story is about the form. When preparing the strategy message, it is important to clearly separate between the strategy, that is, the target state, and the implementation, that is the path to the target state. Otherwise, receivers of the message may fall into the so-called *lost-in-translation trap*.

### 13.6.1   The Traditional Strategy Message

Traditional strategy communication focuses on five key elements:

(1)   A *mission statement*, describing the firm's purpose and identifying the scope of its operations.
(2)   A *vision statement* describing the firm's high-level objectives and how it wants to be perceived.
(3)   A *set of values* that describe the believes on which the vision and mission statements are based.
(4)   A *prose description* of the key aspects that will lead to success.
(5)   A set of quantified *key performance* indicators (KPI) that a firm wants to achieve and that help measure success.

The main challenge with this approach to communication is that it is abstract and subject to interpretation. Many managers and executives have a hard time understanding what these statements, especially the mission, vision, and value statements, mean for their decision-making process. This leads to the primary focus being put onto the tangible KPIs. But KPI-based approaches fall short of addressing the challenges faced in a changing environment. They even sometimes lead to managerial decisions aiming at gaming the strategy, focusing on achieving specific KPIs rather than implementing the strategy in its entirety.

If choosing to rely on the traditional strategy message structure, that is, a mission-vision-values based approach, it should be relegated the end of the communication process, that is, as a summary, rather than beginning the strategy message with it.

### 13.6.2   Crafting the Strategy Message in a Design Thinking World

Crafting a successful strategy message relates to using the insights gained during the strategy design process by answering seven key questions while taking a stakeholder perspective:

(1)   Who are the *targeted customer* segments and who is *explicitly out of scope*?
(2)   What *value propositions* will the firm offer to their targeted customers?
(3)   Why should customers choose to do business with the firm, rather than with its competitors? What are the firm's *distinguishing characteristics* that are hard to copy?
(4)   Why will the firm be able to live up to the promises made to its customers? Which *unique capabilities* (resources, knowledge, capital) does the firm possess and exploit?
(5)   What will *change* with the new or revised strategy and what will *remain the same* over time? How does the firm ensure that existing customers, which remain in scope, are not lost?

(6) Why will the strategy allow *generating and/or increasing profits over time*? How is the financial viability ensured?

(7) Why are the goals set with the new or modified strategy *realistically achievable*?

The answers to these questions should be short and concise. There must be coherence among the answers. The answers should be such that they can be used as a decision support tool during strategy implementation.

The answers to the seven strategy message questions may be used to define the firm's mission (based on answers to questions 1 and 2), its vision (based on answers to questions 2 and 3), and its values (focusing on answers 3 and 4). The mission-vision-values statements must never be perceived as a replacement for the overall strategy message.

**Example** Table 13.1 illustrates possible answers to the strategy message questions for a manufacturing firm competing through innovation, that is, implementing a differentiating strategy along the offerings strategic focus, in the automotive parts supplier industry, focusing on the affected employees' audience.

**Table 13.1** Typical answers to the seven strategy message questions for an innovative automotive parts supplier focusing on the affected employees' audience

| Question | Answer |
| --- | --- |
| (1) Who are the targeted customer segments and who is explicitly out of scope? | − We target the ten largest car manufacturers globally, allowing them to be innovative by using our offerings<br>− We will only work with those manufacturers that also want to be perceived as being innovative and invest significant capital in innovation |
| (2) What value proposition will the firm offer to its customers? | We will offer, at any given point in time, the most innovative automotive parts for being used in new cars, focusing on safety and comfort, as primary innovation drivers |
| (3) Why should customers choose to do business with the firm, rather than with its competitors? | We have shown over the years that we can be one of the most innovative companies in the automotive parts industry<br>Going forward, we will explicitly focus on safety and comfort and collaborate with customers to help them be innovative by using our innovative offerings |
| (4) Why will the firm be able to live up to the promises made to its customers? | − We will collaborate with universities active in research on automotive safety and comfort. By doing so, we ensure being able to identify innovative technologies early on<br>− We will continue to implement agile processes ensuring that academic insights can quickly be translated into innovative automotive parts, supporting innovation for the car buyer |

(continued)

**Table 13.1**  (continued)

| Question | Answer |
|---|---|
| | − We will heavily invest in research and development around safety and comfort, that is, at least 5% of our revenues |
| (5) What will change with the new or revised strategy and what will remain the same over time? | − We will focus our offering on innovations that support safety and user comfort. We move away from quantity towards quality of products offered<br>− We will increase our collaboration with car manufacturers to help them get the most out of our innovations. To do so, we will introduce agile product development and roll-out processes<br>− Out typical product lifespan will be reduced to around 3–5 years, from the current 10–20 years |
| (6) Why will the strategy allow generating and/or increasing profits over time? | By being perceived as an innovative part supplier that puts the innovation capabilities of the car manufacturers at the forefront, we will be able to build lasting relationships that are valued higher than automotive parts providers that primarily compete on price |
| (7) Why are the goals set with the new or modified strategy realistically achievable? | We have shown in the past that we are able to design and deliver high quality innovative products. By taking a more focused approach and implementing a closer collaboration with our customers, chances are increased that our products are valued by our customers over offerings from competitors. In addition focusing on safety and quality, will allow reducing costs by limiting the number of products offered |

## 13.7  Telling the Story

Once the content of the strategy message has been defined by answering the seven key questions, it must be put in a form that is tailored to the target audience. Depending on the audience, more or less details may be included. It is important to get the trade-off between qualitative and quantitative statements right. For example, employees focus on understanding the meaning of the strategy for themselves, whereas analysts and investors prefer quantified and comparable assertions. In all cases, the message should be explicit and compelling.

Storytelling must translate the key properties of the firm's strategy into a compelling and accessible narrative that connects the past with the future in a cohesive way (Mootee 2013). Telling the strategy message story focuses on the three dimensions *inspiring*, *educating*, and *reinforcing*. To do so, it may either rely on *quotes* or on *metaphors*. One possible way of structuring the storyline is to follow the five steps in process K.6.

**Process K.6—Telling the Story**

K.6.1 Setting the stage by creating a burning platform which describes the challenges faced

K.6.2 Describing where the firm wants to be with respect to

- customers,
- offerings,
- capabilities,
- financials, and
- competitors

K.6.3 Explaining why the strategy as described is

- desirable,
- feasible,
- viable, and
- distinct

K.6.4 Illustrating what is different this time and why this will lead to a superior outcome

K.6.5 Finishing the storytelling by focusing on what is in it for the audience targeted by the message

Depending on the communication channels used, different media may be employed to support the delivery and customization of the storytelling.

According to Mootee (2013), a great strategy message story must show seven characteristics:

(1) It must be *collaborative*, engaging multiple stakeholders in sharpening the narrative and delivering the message.
(2) It must be *engaging*, taking the audience on a journey to the future.
(3) It must be *structured*, allowing the audience to follow the reasoning behind the strategy message.
(4) It must be *performative*, using multiple media and relying on dramatic techniques, including tempo and timing.

(5) It must be *tangible*, illustrating the strategy message using prototypes, case studies, or demonstrations.
(6) It must be *fun*, engaging the audience into the message delivery, for example, through workshops, games, or simulations.
(7) It must be *real*, focusing on plausibility and applicability by operationalizing abstract concepts.

The story behind the strategy message must establish a purpose and connect with the audience.

## 13.8   Validating that the Strategy Message is Understood

Similar to the validating step of the business model layer (process V, Chap. 11), the strategy message and its delivery need to be tested before being rolled-out on a large scale. It is important to ensure that the message is understood as intended. To do so, the five-stage approach described by process K.7 should be used.

### Process K.7—Validating the Strategy Message

K.7.1   Identifying a test audience by sampling a representative subset of the targeted stakeholders (a mock-up population that slips into the role of the target audience may be used, if required by confidentiality constraints)
K.7.2   Defining how to measure success of the strategy communication process
K.7.3   Delivering the crafted strategy message to the target test audience
K.7.4   Measuring the success of the strategy message delivery and learning from the feedback received (requiring the test audience to play-back the strategy message or answering a pre-defined set of questions regarding the strategy may be used)
K.7.5   Adjusting the strategy message and its delivery, if needed, based on the results of step K7.4 and iterating to step K.7.1 (ideally the target test audience should be different during each iteration)

## References

Jones, P. (2008). *Communicating strategy*. Hampshire, England: Gower Publishing.
Mootee, I. (2013). *Design thinking for strategic innovation*. Hoboken, NJ: Wiley.

# Index

**A**
Abductive approach, 11
Abductive reasoning, 10, 15, 16, 20, 63
Absolute viewpoint, 182
A/B testing, 174
Access to capital, 99
Activity, 135, 191
Adjacent industry, 60
Affected employee, 204
Aggregating multiple prototypes, 162
Ambiguity, 53
Analogy, 111
Analyst, 126
Analytical thinking, 19
Anti-conventional thinking, 157
Associating, 138
Association, 111
Assumption, 168, 171
Assumption validating process, 166

**B**
Bargaining power, 189
Behavioral pattern, 118
Being different, 183
Being superior, 183
Big bang approach, 208
Blockbuster, 3
Borrowing challenge, 54
Brainstorming, 156
British design council, 23
Budget, 54
Business model, 7, 29
Business model canvas, 12
Business model layer, 64, 68, 111
Buyer driven industry, 100

**C**
Calibrating, 138
Capabilities, 30, 32, 33, 63, 98, 135

Capacity constraint strategy, 190
Capital availability, 99
Capital resources, 40
Change capacity, 56
Chatham House Rule, 126
Choice, 153
Choices made are interdependent, 73
Close-end question, 174
Coach, 53
Cognitive school, 8
Collaborative, 213
Commodity industry, 99
Commodity offering, 184
Commonality, 162
Common language, 30, 132
Communicating process, 74
Communication, 201
Communication channels, 152
Compacting, 151
Competing layer, 64
Competing on differentiation, 32, 182
Competing on price, 32
Competing process, 73
Competition game, 196
Competition on price, 182
Competitive advantage, 5, 93, 191, 192
Competitive environment, 181
Competitive equilibrium, 182, 197
Competitive position, 7, 63, 73
Competitor, 188
Complementarity, 162
Complementor, 188
Complexity of validation, 147
Configurational school, 9
Confirmatory interview, 174
Consideration set, 133
Consistency, 162
Continuity over time, 73, 192
Contradiction, 111

© Springer Nature Switzerland AG 2020
C. Diderich, *Design Thinking for Strategy*, Management for Professionals,
https://doi.org/10.1007/978-3-030-25875-7

Contrast question, 124
Contribution to success, 147
Convergent thinking, 25
Convincing, 201, 203
Core business, 60
Core capability, 84
Core industry, 60
Cost advantage, 40
Cost structure, 42
Create, 24
Creative, 52
Creative people, 145
Creative problem solving, 21
Cultural school, 8
Current detailed business model, 112
Customer, 30, 32, 33, 97, 186, 205
Customer centric competitive advantage, 193
Customer decision journey, 33, 133
Customer delivery, 38
Customer jobs-to-be-done, 37
Customer need, 93
Customer relationship, 38
Customer segments, 37

**D**
D.School, 23
Darden School, 24
Decision owner, 205
Decision support tool, 166
Decision taker, 52, 152
Declining and failing element, 112
Decomposing, 151
Deferred judgment, 156
Defined, 23
Deliver, 23
Delivery, 152
Delivery experience, 133
Demographic, 117
Descriptive question, 124
Descriptive school, 8
Design culture, 22
Designing, 26, 94
Designing process, 71, 147
Design school, 7
Design thinking, 11, 15
Design Thinking for Strategy (DTS), 26, 64
Desirability, 84, 147, 160, 176
Desirable, 12
Detailed business model, 35, 64
Develop, 23, 24
Different, 96
Differentiation, 100, 183
Differentiation advantage, 40
Discounter strategy, 99

Discovery, 23
Disruptive strategy, 98
Disruptor firm, 113
Distinct value proposition, 73
Divergent thinking, 25
3-d model, 158
3-d print of the mind, 158
Domain, 120
DuPont tree, 136

**E**
Ecological, 89
Economic, 87
Ecosystem, 15
Educating, 213
Element-based assumption, 168
Eliminating risk, 165
Emotional decision factor, 184, 194
Empathize, 23
Employee, 188
Engaging, 201, 213
Entrepreneurial school, 8
Environmental analysis, 80
Environmental school, 9
Equilibrium state, 197
Equilibrium theory, 197
Ethnographic approach, 115
Ethnographic interview, 124
Ethnographic observation, 120
Example, 17
Experiment, 171
Explore, 24
External audience, 204
External expertise, 54
Externality-based assumption, 168
Extreme informant, 116

**F**
Facilitator, 53
Failing fast to succeed faster, 72
Fast follower, 99
Feasibility, 84, 147, 160, 177
Feasible, 12
Figures of speech, 158
Financials, 30, 32, 33, 99
First idea, 151
First mover, 98
Five questions on strategy, 64
Five why tool, 125
Focus group, 126
Forgetting challenge, 54
Formulating abstraction, 138
Formulating assumption, 141
Forward thinking, 52

Foundation layer, 64
Framework, 132
Framing the foundation, 79
Frequency of payment, 155
Fun, 214
Fund, 54
Future value creation, 103

**G**
Gain, 98
Game, 197
Game theory, 197
Game tree theory, 197, 199
Geographic, 117
Grand-tour, 120
Group perspective, 126
Guiding principle, 49

**H**
Hasso Plattner Institute, 24
High-stake game, 181
Historical data, 166
Humanity culture, 22
Hypothesis testing, 165

**I**
Ideate, 23, 24
Ideating, 23
Identifying interdependencies, 138
IDEO, 22
Implement, 24
Implementer, 205
Implementing, 23
In-depth observation, 120
Industry trend, 67
Influencer, 205
Informant, 115, 126
Information overload, 122
Informing, 201
Innovation, 98
Innovation culture, 53
Inspiring, 23, 213
Internal audience, 204
Internal resource, 54
Interpreting, 131
Intuition, 20
Invention, 33, 98
Inverting, 151
Investor, 205
Iterative, 16
Iterative approach, 208

**J**
Jobs-to-be-done, 33, 80, 93, 97

Journey map, 33, 134
Judgmental validation, 166

**K**
Key activities, 40
Key performance indicator, 210
Key players, 135
Key resources, 40
Knowledge, 138

**L**
Labor resources, 41
Lead, 133
Learning, 26, 94, 131
Learning challenge, 54
Learning process, 69
Learning school, 8
Legal, 88
LEGO® SERIOUS PLAY® method, 158
Level of abstraction, 30
Lightweight business model, 31
Listening, 111
Lost-in-translation trap, 209

**M**
Magnifying, 151
Making these choices transparent, 209
Manager, 204
Managing cost, 99
Managing the uncertainty, 166
Marginal added knowledge, 166
Market squeezing strategy, 190
Mature firm, 112
Measurement criterion, 171
Media, 205
Mental model, 131
Metaphor, 158, 213
Mind game, 157
Mini-tour, 120
Min-max approach, 197
Mission statement, 210
MIT, 24
Mock-up, 173
Moderator, 126
Monopolistic strategy, 190
Multiplying, 151
Multi-variate testing, 174

**N**
Narrative, 158
Nash equilibrium, 197
Need, 97
Netflix, 3
Networker, 205

New jobs-to-be-done, 154
New target population, 154
Niche segment, 182
Non-cooperative game, 197
Non-customer segment, 82
Not affected employee, 204
Not invented here, 53

**O**
Observing, 24, 26, 94, 115
Observing process, 68, 115
Offering, 30, 32, 33, 98
Outsourced activities, 40
Over-analysis fallacy, 79
Owner of the strategy message, 203

**P**
Pain, 97
Participatory design, 20
Passively observing, 119, 121
Perceived trust, 155
Perception of cost, 193
Performance, 184
Performative, 213
Perishable resources, 40
Persona, 33, 96, 115
Personas framework, 116
PESTLE framework, 87
PESTLEWeb, 89
Planning school, 8
Political, 87
Political school, 8
Positioning school, 8
Prescriptive school, 7
Price elasticity strategy, 190
Pricing model, 99, 154
Prioritizing, 150
Process manager, 53
Process moderator, 53
Process supporter, 53
Products and Services, 39
Prototype, 16, 23, 24, 173
Psychographic characteristics, 118
Purchase decision, 133

**Q**
Quantifiable hypothesis, 166
Quotes, 213

**R**
Rational decision factors, 193
Real, 214
Reassuring, 201, 203
Recombining existing capabilities, 154

Re-configuring existing insights, 114
Reducing risk, 165
Reference point, 137
Refine, 24
Reflect, 24
Regulator, 205
Reinforcing, 213
Relationship-based assumption, 168
Relative viewpoint, 182
Rephrasing, 138
Resource-based approach, 93
Resources, 5, 63, 98
Revenue generating mechanism, 84
Revenue streams, 41
Root cause analysis, 23

**S**
Saboteur, 204, 205
Scheduling, 54
Science culture, 22
Secondary research, 127
Selling more, 154
Sense making, 131
Sense of urgency, 209
Set of values, 210
Setting the stage for change, 201
Showing seriousness, 203
Signaling strategy, 190
Skill, 98, 154
Skill resources, 41
Societal, 88
Split testing, 174
Stakeholder map, 50
Stakeholder, 13, 19
Stanford University, 23
Start-up firm, 112
Statistical hypothesis testing, 103
Strategic fit, 191
Strategic focus, 7, 32, 64, 68, 93
Strategic value discipline, 32, 93
Strategy, 4
Strategy analysis tool, 93
Strategy brief, 49
Strategy designer, 52
Strategy design process, 10, 63
Strategy design theory, 63
Strategy experiment, 103
Strategy for competing, 190
Strategy hypothesis, 71, 103
Strategy is about choice, 191
Strategy team, 52
Stretching, 151
Strong collaboration, 145
Structured, 213

Structured question, 124
Stuck in the middle trap, 97
Substitute, 153
Superior, 96
Superiority, 100, 184
Supplier, 189
Support, 153
Supporter, 204
Survey, 175
SWOT, 5, 8
SWOT analysis, 85, 135
Synthesize knowledge, 131

**T**
Tacit knowledge, 22
Tailored value chain, 73
Tangible, 214
Target audience, 201
Target industry, 49, 59
Target informant population, 171
Target population, 111, 115, 149
Technological, 88
Temporary failure, 53
Test, 23, 24
Thinking aloud, 122
Threshold, 171
Timeline, 54
Timing of payment, 155
Trade-off, 73, 191
Types of jobs-to-be-done, 153

**U**
Unbiased moderation, 159

Uncertainty, 53
Understand, 24
Understanding the environment, 79
Union, 205
Unique, 12, 183, 184
Units of value, 155
University of Potsdam, 24
University of Virginia, 24
Usage, 153

**V**
Validating, 16, 26, 94
Validation phase, 165
Validation process, 71
Value chain, 5, 135
Value characteristic, 193
Value proposition, 39, 84
Viability, 84, 147, 160, 177
Viable, 12
Vision statement, 210

**W**
What if, 24
What is, 24
What works, 25
What wows, 24
Wicked problem, 10
Willingness to pay, 146
Working capital requirement, 155

**Z**
Zero-sum game, 197